Biophilic School Design for Health and Wellbeing

Biophilic School Design for Health and Wellbeing

Guest Editors

Rokhshid Ghaziani
Kenn Fisher

MDPI

Basel • Beijing • Wuhan • Barcelona • Belgrade • Novi Sad • Cluj • Manchester

Guest Editors
Rokhshid Ghaziani
School of Architecture, Art and Design
University of Portsmouth
Portsmouth
UK

Kenn Fisher
Faculty of Architecture, Building and Planning
University of Melbourne
Melbourne
Australia

Editorial Office
MDPI AG
Grosspeteranlage 5
4052 Basel, Switzerland

This is a reprint of the Special Issue, published open access by the journal *Architecture* (ISSN 2673-8945), freely accessible at: https://www.mdpi.com/journal/architecture/special_issues/QR208XZG01.

For citation purposes, cite each article independently as indicated on the article page online and as indicated below:

Lastname, A.A.; Lastname, B.B. Article Title. *Journal Name* **Year**, *Volume Number*, Page Range.

ISBN 978-3-7258-5533-9 (Hbk)
ISBN 978-3-7258-5534-6 (PDF)
https://doi.org/10.3390/books978-3-7258-5534-6

Cover image courtesy of Rokhshid Ghaziani

Contents

About the Editors

Rokhshid Ghaziani

Rokhshid Ghaziani is a Senior Lecturer in Architecture and Interiors at School of Architecture, Art & Design, the University of Portsmouth, and is a Fellow of the Higher Education Academy. She joined the University of Portsmouth in 2022, following five years of teaching and research at De Montfort University. Since 2005, Rokhshid has taught a wide range of subjects for Architecture and Interior Design courses at various Schools of Architecture and Design in the UK. Rokhshid has a Doctorate in Architecture, awarded by the University of Sheffield in 2009. Her research and teaching combine interests in the design of schools, participatory design with children, and biophilic design for health and wellbeing. Since 2019, she has led a number of funded research projects in the UK and contributed to a GCRF Networking project in collaboration with Indonesia, Thailand, and Malaysia.

Kenn Fisher

Kenn Fisher has practised as a 50% research/teaching academic and 50% learning and health environment consultant, with his research informing practice and vice versa through a translational design process. This integrates an evidence-based research approach into environmental design practice. His multi-disciplinary experience ranges across all educational levels, as well as medical and health sectors. His roles have included teaching and research; consulting in functional design briefing and campus planning; and project, facility, and design management. Kenn was Head of the OECD Centre for Effective Learning Environments in Paris in 1997/8 and has consulted with many educational sectors in the Middle East, Europe, Australia, New Zealand, Hong Kong, China, Malaysia, UK, and the USA. He is a co-founder of the Learning Environments Applied Research Network at the University of Melbourne, with the Centre winning some USD 7 million in research grants since its inception in 2009. His recent research focus is on evidence-based biophilic urbanism and biophilic design for health and wellbeing.

architecture

MDPI

Editorial

Biophilic School Design for Health and Wellbeing

Rokhshid Ghaziani [1,*] and Kenn Fisher [2,*]

1 School of Architecture, Art and Design, University of Portsmouth, Portsmouth PO1 2UP, UK
2 Melbourne School of Design, University of Melbourne, Parkville, VIC 3010, Australia
* Correspondence: rokhshid.ghaziani@port.ac.uk (R.G.); fisherk@unimelb.edu.au (K.F.)

1. Motivations for This Reprint

This Reprint title—'Biophilic School Design for Health and Wellbeing'—suggests the coverage of a range of factors through which it is contended that key elements of nature and the built environment have a direct link to the health and wellbeing of students and teachers.

In short, this is an extension of the much researched 'sick building syndrome' which was examined in depth in 1983 by the World Health Organisation [1]. It is, therefore, not surprising that the concept of biophilia also re-emerged around this time through E. O. Wilson [2]. The original enunciation of the hypothesis was developed in Eric Fromm's [3] first use of the term in his publication 'The Anatomy of Human Destructiveness'. Fromm suggested that humans have a deep connection to nature in part to ensure a healthy life, but also as a nurturing resource.

Fromm labelled this connection 'biophilia' or the love of nature. Later, following Kellert's [4] work, this connection has been confirmed in a myriad of studies, with van den Bosch's [5] 'Nature and Public Health' being one of the most comprehensive in its coverage. This work came into sharp focus in 2020 with the advent of the global pandemic COVID-19. This emergence resonates with the Tuberculosis pandemic in the late 1890s and early 1900s. At that time, the 'open air school movement' emerged as a potential counter to this disease. In some respects, the 21st Century biophilic school design campaign reprises that movement [6].

2. Aim and Purpose

Thus, a prefix to the title might well be 'Evidence-Based', with the ten articles in this Reprint sourcing evidence-based peer reviewed literature, which informs the planning and designing methodology through biophilic principles towards enhancing the health and wellbeing of students and teachers. This evidence-based approach parallels those used by health planners in designing hospitals, clinics, and allied health facilities.

Health planners have adopted an evidence-based method by leveraging the approach used in medicine and public health, where all outcomes are measured through an evidence-based lens. Such an approach has not yet been applied to the planning and design of school facilities with any significant rigour. We note, however, that data led teaching and learning approaches are becoming the norm in education, but are yet to feature significantly in the design of learning environments. Many studies are now making the connection between the built and natural environment and health and wellbeing through Terrapin's [7] 14 Biophilic Design Patterns.

https://doi.org/10.3390/architecture5030065

3. Subject and Scope

The Reprint covers three key areas: (1) the emerging evidence-base around nature and its impact on health and wellbeing; (2) using these findings to inform biophilic design principles for schools; and (3) the use of selected school and university case studies to illustrate how these two intersect in practice.

The evidence based around nature and its relationship to health and wellbeing explores how biodiversity on the campus grounds, and ultimately within the learning spaces as well, can enhance health indicators and outcomes. Furthermore, wellbeing indicators are also implicated in connection with various natural settings. The emerging biophilic principles can be directly related to these natural elements and the measurement of various human factors in a variety of settings is possible and presented.

These biophilic principles include overall greening targets, including the use of school grounds to improve and increase natural and biodiversity affordances, the introduction of more natural elements into school architectural design and also to enhance indoor environmental quality performance.

The first of the three parts, namely 'the emerging evidence-base around nature and its impact on health and wellbeing' is covered by three authors. Tim Baber & Ben Cleveland [Contribution 1] explore the interaction of biophilia and adolescents' sense of place in Australian vertical schools Kenn Fisher [Contribution 2] investigates the concept of the biophilic school through carrying out a critical synthesis of evidence-based systematic literature reviews. Finally, Fiona Gray & Andrea Downie [Contribution 3] consider the design of thriving school ecosystems: the synergy of biophilic design, wellbeing science, and systems science.

The second part uses these findings to 'inform biophilic design principles for schools'. Rokhshid Ghaziani [Contribution 4] suggests the need for re-thinking biophilic design for primary schools in exploring children's preferences. Bethania Lanzaro and Marcella Ucci [Contribution 5] probe teacher and parent perceptions of biophilic conditions in primary school environments and their impact on children's wellbeing whilst Kari Leif & Vivian Loftness [Contribution 6] develop a toolkit of biophilic interventions for existing schools to enhance student and faculty health and performance.

The third part makes use of 'selected school and university case studies to illustrate how these two intersect in practice'. William Browning & Jim Determan [Contribution 7] conduct a study of a secondary school to determine learning outcomes of the biophilic design of schools. Alan J Duffy [Contribution 8] investigates the 'nature' of vertical school design through a biomimicry lens. Elia Ebrahimi Salari & Nigel Westbrook [Contribution 9] focus on wellbeing through mediated spaces including informal learning and physical activity environments in vertical schools. Yuqing He, Jacky Bowring, & Gillian Lawson [Contribution 10] explore mental health improvements through campus landscape design: insights from New Zealand universities.

4. The Suggested Audience

The primary audience is school principals, as they are in the best position to lead such an approach to health and wellbeing on their campuses. In turn, they can lead and calibrate the staged incorporation of these features with teachers and in concert with budget availability.

But so too are many other participants in this approach and campaign. This includes facilities managers within Ministries of Education, School Principal Associations, Subject Teacher Associations, Architects, Engineers, and other professionals involved in the planning, design, construction and management of school facilities. Furthermore, health and wellbeing advocates in the abovementioned sub-sectors are key to the implementation of this approach.

Another key sector which is often overlooked is faculties of education within universities. Teacher trainees are rarely exposed to the physical environment as a pedagogical affordance and, in particular, the learning environment as a factor in influencing the health and wellbeing of both teachers and students alike.

It is hoped that this Reprint summarises key aspects of an evidence-based biophilic school design approach to health and wellbeing in a way that provides effective evidence and the ways and means to implement such a strategy.

Author Contributions: Writing—original draft preparation, R.G. and K.F.; writing—review and editing, R.G. and K.F. All authors have read and agreed to the published version of the manuscript.

Conflicts of Interest: The authors declare no conflicts of interest.

List of Contributions:

1. Baber, T.; Cleveland, B. Biophilia and Adolescents' Sense of Place in Australian Vertical Schools. *Architecture* **2024**, *4*, 668–691. https://doi.org/10.3390/architecture4030035.
2. Fisher, K. The Biophilic School: A Critical Synthesis of Evidence-Based Systematic Literature Reviews. *Architecture* **2024**, *4*, 457–478. https://doi.org/10.3390/architecture4030025.
3. Gray, F.; Downie, A. Designing Thriving School Ecosystems: The Synergy of Biophilic Design, Wellbeing Science, and Systems Science. *Architecture* **2024**, *4*, 594–612. https://doi.org/10.3390/architecture4030031.
4. Ghaziani, R. Re-Thinking Biophilic Design for Primary Schools: Exploring Children's Preferences. *Architecture* **2025**, *5*, 42. https://doi.org/10.3390/architecture5030042.
5. Lanzaro, B.; Ucci, M. Teacher and parent perception of biophilic conditions in primary school environments and their impact on children's wellbeing. *Architecture* **2024**, *4*, 367–389. https://doi.org/10.3390/architecture4020021.
6. Leif, K.; Loftness, V. A Toolkit of Biophilic Interventions for Existing Schools to Enhance Student and Faculty Health and Performance. *Architecture* **2024**, *4*, 445–456. https://doi.org/10.3390/architecture4020024.
7. Browning, W.; Determan, J. Outcomes of Biophilic Schools. *Architecture* **2024**, *4*, 479–492. https://doi.org/10.3390/architecture4030026.
8. Duffy, A. The 'Nature' of Vertical School Design—An Evolving Concept. *Architecture* **2024**, *4*, 730–744. https://doi.org/10.3390/architecture4030038.
9. Ebrahimi Salari, E.; Westbrook, N. Ascending to Well-Being through Mediated Spaces: An Alternative to Informal Learning and Physical Activity Environments in Vertical Schools. *Architecture* **2024**, *4*, 613–638. https://doi.org/10.3390/architecture4030032.
10. He, Y.; Bowring, H.; Lawson, G. Promoting Mental Health Through Campus Landscape Design: insights from New Zealand Universities. *Architecture* **2025**, *5*, 16. https://doi.org/10.3390/architecture5010016.

References

1. European Centre for Environment and Health, WHO. *WHO Guidelines for Indoor Air Quality: Selected Pollutants (PDF)*; EURO Reports and Studies No. 78; Bonn Germany Office, WHO Regional Office for Europe: Copenhagen, Denmark, 2010.
2. Wilson, E. *Biophilia*; Harvard Publishing: Boston, MA, USA, 1984.
3. Fromm, E. *The Anatomy of Human Destructiveness*; Holt, Rinehart and Winston: New York, NY, USA, 1973.
4. Kellert, S. *Nature by Design: The Practice of Biophilic Design*; Yale University Press: New Haven, CT, USA, 2018.
5. van den Bosch, M.; Bird, W. (Eds.) *Nature and Public Health: The Role of Nature in Improving the Health of a Population*; Oxford University Press: Oxford, UK, 2018.

6. Kue, M.; Barnes, M.; Jordan, C. Do Experiences with Nature Promote Learning? Converging Evidence of a Cause-and-Effect Relationship. *Front. Psychol.* **2019**, *10*, 305. [CrossRef] [PubMed]
7. Terrapin Bright Green. 14 Patterns of Biophilic Design Improving Health & Well-Being in the Built Environment. Available online: https://www.terrapinbrightgreen.com/wp-content/uploads/2014/09/14-Patterns-of-Biophilic-Design-Terrapin-2014p.pdf (accessed on 7 August 2023).

architecture

MDPI

Article

Biophilia and Adolescents' Sense of Place in Australian Vertical Schools

Tim Baber * and Benjamin Cleveland

Learning Environments Applied Research Network (LEaRN), Faculty of Architecture, Building and Planning, University of Melbourne, Parkville, VIC 3010, Australia; benjamin.cleveland@unimelb.edu.au
* Correspondence: tbaber@student.unimelb.edu.au

Abstract: This article describes the place relationships adolescents have with natural elements in recently developed 'vertical schools' in Australia: a relatively new school typology in the country, generally considered to be over four storeys high. Vertical schools are being built in Australia's largest cities in response to the need for new schools where land is scarce. Drawing on qualitative research into the place relationships that adolescents have with their learning environments in two Australian vertical schools, this article explores the ways young people seek closer associations with nature in multi-storey educational settings. The research adopted a phenomenological approach to 'place research', asking what makes a place a place? Further, it recognised that places have the capacity to shape the 'becoming' of a person, that 'becoming' happens when there is a certain resonance between a place's cycles, and that identity is created out of difference. As such, 'place as complex adaptive assemblage theory' was employed to study the interplay between location and experience in selected vertical schools, also drawing on the earlier theoretical work of Relph (1976), who suggested that the essence of a place lies in the unselfconscious intentionality that defines places as profound centres of human existence. Photovoice was employed as a method to elicit insights into the relationships adolescents have with their school as place. Students created photo essays to document their place relationships, revealing deeply insightful personal reflections on their school as place and the meanings they associate with it. Specifically, the article reports on what thirteen adolescent students communicated about the biophilic experiences afforded by their vertical school, including how it influenced their 'becoming'. The findings demonstrated that the participating students not only valued relationships with natural elements, but actively sought biophilic experiences on a regular basis. The desire for more frequent and more significant interactions with natural materials and cycles was common across most students, indicating that the relatively limited biophilic experiences available to students in vertical schools is a challenge that should be addressed in future multi-level educational environments.

Keywords: biophilia; adolescents; vertical schools; place; assemblage; becoming

1. Introduction

This article reports findings from a Ph.D. study into the place relationships of thirteen adolescent students at two Australian vertical schools, with a focus on what they communicated about their relationships with the biophilic elements in and around their schools.

Historically, Australian schools have adopted a horizontal form, featuring generous outdoor environments that have included playing fields and courts, pathways, garden beds, courtyards, seating nooks, and swimming pools in some schools. These outdoor areas have afforded students the opportunity to encounter plants and other natural elements, including large established trees and shrubs that reflect changing seasons. Many primary schools have vegetable patches and kitchen-garden clubs where students, parents/carers, and other school community members can grow fresh produce, including fruits and vegetables, and prepare meals. These ground-level settings also offer exposure to weather events, including sunny and rainy days, especially on campuses featuring distributed buildings.

By contrast, newly developed vertical schools on small parcels of land tend to offer limited outdoor space and experiences, and therefore fewer opportunities for contact with nature. Vertical schools have been emerging in Australia since 2018 in response to the need for new schools in the country's largest cities. For example, Sydney is under acute pressure to provide more inner-city schools. The Greater Sydney Commission [1] predicts that the number of children in the city will increase by 41% by 2039. Across the country, it is predicted that an additional 700 new schools, not just vertical, will be required between 2016 and 2036 to accommodate an additional one million school-aged students [2,3]. In response, the country's education authorities have begun building larger schools on smaller sites, especially in inner-city locations.

Vertical schools are tall—generally considered to be over four storeys—but are not defined by this attribute alone. Carroli et al. [4] argued that the morphology of this 'new' school typology can best be understood in relation to the building's height, its spatial configuration and use, and its urban context. With respect to height, Taylor [5] suggested that schools taller than six storeys should be considered high-rise, while Newton [6] delineated between mid-rise vertical schools (four to seven storeys) and high-rise vertical schools (including up to seventeen storeys in the Australian context).

The following research question was investigated: How do adolescent students experience a sense of place in Australian vertical schools? Seeking understandings of students' experiences of place in relation to their schools' physical attributes, both biophilic and non-biophilic, was the overall objective of the research, with a particular orientation towards how these experiences may impact identity formation and becoming. The findings offer insights into students' place experiences, the values they associate with their school as place, and the types of relationships that exist between students and their school's physical attributes. Amidst other project findings, this article narrows the focus to report specifically on students' biophilic place relationships at their schools, following the strong emergence of biophilic relationships in the data.

2. Biophilic Design and Schools

2.1. Biophilia

Erich Fromm, a German American social psychologist, developed the term 'biophilia' to describe a "passionate love of life and all that is alive" [7] (p. 525). Subsequently, American biologist Edward Wilson [8] argued that human beings have a natural affinity for life and biophilia is the essence of our humanity, binding us to other living things. More recently, the term has come to refer explicitly to people's relationships with nature. Indeed, Kellert [9] (p. 3) defined biophilia as "the inherent human inclination to affiliate with natural systems and processes, especially life and life-like features of the nonhuman environment".

2.2. The Biophilious Person

The origins of the term biophilia lie in Fromm's desire to describe the antithesis of death. He contrasted Freud's 'death instinct' and the necrophilous character (e.g., Stalin, Hitler, and Himmler) with its opposite, 'life instinct'. Fromm [7] (p. 525) argued that life instinct is the "effort of Eros to combine organic substance into ever larger unities, whereas the death instinct tries to separate and to disintegrate living structure". He proposed that the 'biophilious person' has a desire for further growth, to construct rather than retain, to be more than to have more, to be capable of wondering, to see something new rather than to seek confirmation in the old, and loves the adventure of living more so than certainty.

2.3. Biophilic Design

Biophilic design attempts to translate people's inherent affinity for natural systems into the built environment [8–10]. Two well-documented approaches to describing and documenting biophilia in the built environment have emerged in the past ten years. Kellert [9] proposed that biophilic design should be considered in relation to two dimensions, an organic or naturalistic dimension, and a second dimension related to people's connections to and sense of place. Alternatively, Ryan and collaborators [11,12] have developed Biophilic Design Patterns and categorised biophilia according to cognitive, physiological, and psychological health responses to it. They identify nature in the space, natural analogues, and nature of the space as three important dimensions of the concept. Approaches such as these have informed the architectural design industry as it has adopted a biophilic sensibility over the past decade.

2.4. Research into Biophilic School Design

Research into the biophilic design of schools has a relatively short history, with academic interest in the topic concentrated within recent years following biophilic design research in other design sectors (e.g., commercial buildings). In 2021, Ghaziani et al. [13] proposed ten biophilic design patterns for schools, arranging them according to nature in the space and natural analogues. In 2022, Watchman and colleagues [14] formulated a visual biophilic design vocabulary for schools, proposing a common graphic language to enable researchers and designers to describe biophilic spaces through shared terms and logic. In 2023, Aminpour [15] found primary children attending vertical schools in Australia have an affinity for biophilic elements across the categories of direct experience of nature, indirect experience of nature, and the nature of the space. Meanwhile, Mohammed et al. [16] highlighted a lack of biophilic features across six preschools in Duhok, Iraq. With only limited research into biophilic school design—including its influence on students and their learning—there remains significant scope for further research in the field.

2.5. The Benefits of Natural Elements in Schools

Although research specifically into biophilic school design is limited, research into the relationships between young people and nature is more extensive, indicating that young people benefit from natural elements in their schools, including:

- The presence of natural features, with young people perceiving the restorative effectiveness of school playgrounds in correlation with the number of trees and volume of vegetation present [17,18].
- Diminished psychological stress at schools that are 'greener' [19].
- Views through windows to green landscapes, increasing student attentiveness and assisting with stress recovery [20].
- Contact with nature at school, promoting cognitive ability and enhancing working memory [21].
- Exposure to natural stimuli, as opposed to urban stimuli, improving working memory [22].

The presence of natural elements may be acutely felt in urban schools, where the relative 'greenness' of the surrounding areas may influence school-wide academic performance [23]. Indeed, Amicone et al. [24] found that students at urban schools who encounter natural environments in their break times return to class with increased sustained and selective attention relative to their counterparts in built-up environments. Dense plantings of trees appear to have the most impact, with tree cover density being positively associated with adolescents' academic performance. Hodson and Sander [25] found that tree cover correlated with better reading scores, when compared with lawns, shrubbery, and water bodies, while Li et al. [26] found tree cover density within a one-mile radius of urban high schools was positively associated with test scores and college readiness. While correlations between biophilic relationships and test scores may be an indirect measure of the influence of natural elements in and around schools, it is interesting to see several studies making such connections.

3. Adolescent Identity Formation, Biophilia, and Place

A key task of adolescence is to develop a coherent sense of self and identity [27]. Through an ever-cycling process of movement from the stable to the new, the adolescent's identity morphs and emerges. Branje and colleagues [28] describe this as a process by which young people come to know who they genuinely are. What emerges is a subjective feeling of self-sameness and continuity over contexts and time. The adolescent becomes aware of their distinctiveness and uniqueness, a coherence and similarity forms regarding who they are across various domains, and there develops a continuity over time and across situations [29].

The places adolescents inhabit impact their emerging identities. Bissell [30] (p. 272) writes about affective relationships, being the development of an affective atmosphere that is perceived and sensed through the body and forms part of 'the ubiquitous backdrop' that affects the ways in which spaces are inhabited. Schools, as places, may play a key role in adolescent identity formation.

4. Connecting Theories to Explore Adolescents' Biophilic Place Relationships

To date, students' biophilic place relationships with schools have not been a topic of known empirical research. However, connections between theories about adolescent identity formation, Fromm's ideas about the biophilious person, and place relationships, such as explored by researchers including Dovey [30–32], provide a fascinating theory base for investigating adolescents' biophilic place relationships.

As outlined above, a desire for growth, to construct rather than retain, to be more than to have more, to be capable of wondering, to see something new rather than to seek confirmation in the old, and a love of adventure are key attributes of Fromm's biophilious person [7] and brings into the frame the process of movement from the stable to the new, as accompanies the emergence of adolescents' identities. Here, further connections may be made to Dovey's [33] ideas about the becoming of places. For Dovey, places become; they are assemblages that stabilise dwelling while also encompassing lines of movement and processes that promote something new. He suggests that places become where zones of order, uniformity, familiarity, and a sense of home are challenged and destabilised, where new movements, along with inventiveness and curiosity, challenge and dissolve established boundaries. Ever-emerging processes of places becoming, he contends, occur in the resonances of tensions between the existing and the new.

In this article, it makes sense to seek understandings of adolescent place relationships in a way that considers the tensions that exist between the existing and stable versus the new and the emerging. A brief introduction to place theory below further explores these ideas as a basis for this research.

5. A Brief Introduction to Place Theory

5.1. Origins of Place

Heidegger's [34] foundational contribution to contemporary thinking about place is his notion of Dasein, meaning 'being there' or 'being in'. For him, 'being' (existence) cannot be thought of simply in terms of occupying space as a container. Rather, there is a connection between a person and a place, a continuity between person and place, that is premised on the requirement of 'being-there'—in place. Following this line of thinking, Relph [35] (p. 43) concluded that:

> The basic meaning of place, its essence, does not therefore come from locations, nor from the trivial functions that places serve, nor from the community that occupies it, nor from superficial or mundane experiences ... The essence of place lies in the largely unselfconscious intentionality that defines places as profound centers of human existence.

Such a phenomenological approach to understanding place is not that interested in the unique attributes of places, nor is it primarily concerned with social forces. Rather, it seeks to "define the essence of human existence as one that is necessarily and importantly 'in-place'" [36] (p. 158).

As Cresswell suggests, "this approach is less concerned with 'places' and more interested in 'Place'" [36] (p. 158). "So, rather than asking what this place or that place is like, the phenomenological approach to place asks what makes a place a place?" [36] (p. 96). Further, Malpas [37] (p. 13) argues that place is "the frame within which experience is to be understood" and is "integral to the very structure and possibility of experience" [37] (p. 31). In summary, Seamon [38] (p. 29) concludes that in this sense "places are spatial-temporal fields that integrate, activate, and interconnect things, people, experiences, meanings and events".

5.2. Place as Complex Adaptive Assemblage

Doreen Massey's paper, "A Global Sense of Place" [39], marked a shift regarding how place might be considered. Her work disrupted the Heideggerian premise that place is pre-given and deep-rooted. Instead, she argued that places are open, progressive, and global; that they are outward-looking, defined by multiple and fluid identities; and their character is derived from connections and interactions, rather than origins and boundaries. Massey's work has been widely cited as a plea for a new conceptualisation of place as open and hybrid, "a product of interconnecting flows—of routes rather than roots" [36] (p. 234).

Aligned with such thinking, a conceptualisation of place as 'assemblage' has evolved over recent years, drawing upon the earlier work of Deleuze and Guattari [40]. Dovey [31–33], along with McFarlane [41], are prominent proponents of assemblage thinking when seeking to understand place relations. They argue assemblage thinking is not so much interested in the formation of a place itself, rather in the interactions and processes—the relations—between the social and the material elements, and the resultant temporalities and possibilities that present themselves as a result. Dovey argues for adopting an approach to understanding place that is "deep seated in everyday life without being deep-rooted in fixed origins . . . grounded in the particularities and practices of everyday life" [33] (p. 30).

Referencing Deleuze and Guattari [40], and DeLanda's [42] interpretation, Dovey [33] outlines a way of thinking about place that seeks to replace Heidegger's 'being-in-the-world' with Deleuzian 'becoming-in-the-world': a more dynamic and open sense of place as a multiplicitous assemblage. He notes that the term 'assemblage' is a translation of the French 'agencement', which is synonymous with the words 'layout', 'arrangement', or 'alignment', and to some degree, 'agency'. Assemblage thinking is relational, and assemblages are dynamic, they change and adapt over time. The flows of life that give them their intensity and sense of place are important. In the context of urban research, Dovey suggests that:

> An assemblage is a whole that is formed from the interconnectivity and flows between constituent parts wherein the identities and functions of both parts and wholes emerge from the flows, alliances, and synergies between them. For Deleuze 'It is never filiations which are important, but alliances, alloys' [43] (p. 69) and 'Don't ask what it means, show how it works' [33] (p. 22).

Dovey [33] explains that assemblage thinking also has connections to the science of emergence, complex adaptive systems, and resilience thinking [44–46]. He argues that:

> "Places can be understood as complex adaptive systems where over time a regime or identity with certain characteristics emerges, settles down and becomes more or less resilient. Resilience is defined as the capacity of the system to adapt to change without crossing a threshold into a new 'regime' or 'identity' [46] (p. 32)" [33] (p. 28).

Resilience in this sense is not the capacity to maintain or return to a single stable state but rather a dynamic capacity to move between a range of adaptive states without crossing a threshold of no return. In this sense, an enduring place identity develops as an emergent regime, sustained by complex sets of relations and adaptive capacities. Dovey [33] has taken complex adaptive systems theory (derived from the physical sciences) and assemblage thinking (which resists any reduction of place to a science) and argues that there are many interconnections between the two and that a 'complex adaptive assemblage' is a useful way of understanding place.

6. Research Design

6.1. Case Study Design

A case study methodology was used to investigate how adolescent students experience a sense of place in Australian vertical schools. As discussed above, this article focusses on their relationships with the biophilic elements in their schools, as emerged in the data. The case study methodology was employed to gain in-depth understandings of a contemporary phenomenon within its context, while also offering scope to expand and generalise theory [47].

6.2. Case Study Sites

Two case schools were recruited: City School and Inner-Suburban High School. Pseudonyms are used for the schools and individual participants to protect their identities. The characteristics of these vertical schools are presented below.

Occupying a nine-storey repurposed office building, City School (Figure 1) is an independent non-denominational school catering to students from Reception to Year 12 (all compulsory school years in Australia). The school is located in Melbourne, Victoria, adjacent to an historic park/gardens, and experiences the city's mild-temperate climate. The school features 1500 square metres of outdoor recreation space across three terraces, in addition to extensive indoor settings. The school opened progressively from 2016 and 2018, and now has an enrolment of 750 students from Reception to Year 12.

Featuring four buildings, the tallest being seven storeys high, Inner-Suburban High School (Figure 2) is in the warm-temperate climate of Brisbane, Queensland. It caters to students from Years 7 to 12 (secondary school years in Australia). The campus design includes five 'learning hubs' centred around a green courtyard. The school opened in 2021 and has an enrolment capacity of 1600 students. At the time of data collection, the school was operating with students in Years 7 to 9 only and had an enrolment of 950 students.

6.3. Biophilic Design of the Case Schools

Two of the three Biophilic Design Patterns identified by Ryan and collaborators [11,12] are apparent at both case schools: 'nature in the space' and 'nature of the space'. Neither offer 'natural analogues'.

At City School, nature in the space is offered via three large outdoor terraces and the neighbouring park. The terraces offer direct access to sunlight, breezes, and extensive views to the park, the sky, and beyond. While the school's building features few plants, students utilise the neighbouring park and its many established trees and seasonal planting beds for the occasional lesson, and during breaks. Almost all rooms of the school feature views of the park. The nature of the space is offered via prospective refuge [48], with the facilities offering a range of nooks from which students can be alone or gather with friends to enjoy the extensive views of the park and city in comfort and safety.

The provision of nature in the space is more immediately apparent at Inner-Suburban High School, which features several large outdoor terraces affording views to the city, a neighbouring park, and Mt Coo-tha in the distance. Large plant beds feature on each terrace, offering students close contact with living shrubs. Some terraces are open to the sky, while others are covered. Those that are covered feature tall ceilings or atriums, promoting an abundance of natural light and the free flow of breezes. The school's buildings are arranged around a large central courtyard containing trees, shrubs, and climbing plants, while the school's perimeter also features trees, shrubs, and climbing plants. The neighbouring historic park features large established trees and fronts the Brisbane River. Many rooms feature views of the park, the city, or Mt Coo-tha. Like City School, Inner-Suburban High School provides opportunities for prospective refuge (nature of the space) through its many nooks that offer views within the school and some beyond.

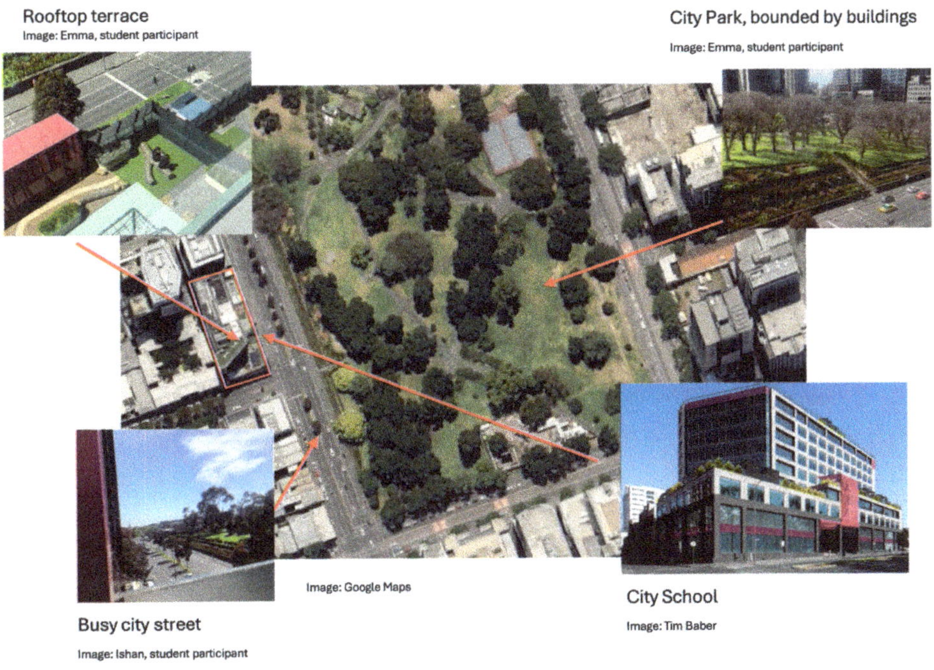

Figure 1. City School [49].

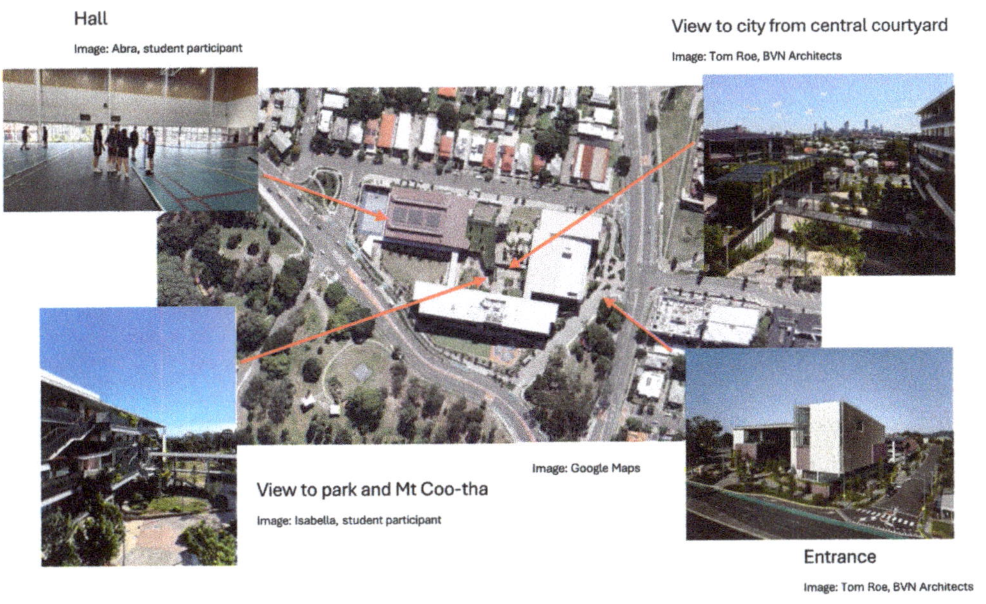

Figure 2. Inner-Suburban High School [50].

7. Data Collection Method

Photovoice was employed to collect data. This method situates cameras in the hands of research participants and invites them to visually represent and communicate their lived experiences to others [51]. Thirteen students from across the two schools were guided through a series of five workshops to document and share photo essays about their place relationships with their schools' environments.

Participants engaged in the following workshops to create their photo essays:

- Workshop One: Photovoice ethics; the power of photography; photographic composition.
- Workshop Two: Developing themes in photography; storyboarding and how to create a photographic essay; accompanied photography at school.
- Workshop Three: Identifying the 'issue'; examples provided of artists and professional photographers who communicate an important issue through a collection of photographs; further storyboarding and emergence of themes; second accompanied photography to refine images.
- Workshop Four: Strengthening and personalising the message through editing and composition.
- Workshop Five: Presentation of photo essays and group discussion, providing the researcher the opportunity to ask clarifying questions.

The data set included each participant's presentation slides and description of their experiences, along with an audio recording and transcript of the accompanying group discussion.

Students were recruited from Years 9 and 10 by calling for volunteers following a school-based presentation of the research project. A total of thirteen students volunteered to participate, including seven from City School and six from Inner-Suburban High School. While not participants, local teachers assisted with logistics, co-hosted workshops, and assisted the Ph.D. researcher to establish rapport with the students.

To conclude the data collection phase in each school, posters of the students' photo essays were exhibited to each school community. Students, parents, community members, and representatives from the relevant education authorities attended the launches of these exhibitions. Accompanying these events, the Ph.D. researcher engaged the participants in feedback sessions where the findings of the entire project were communicated back to the participating students, seeking confirmation and/or additional feedback on what had been found.

8. Analysis

Narrative analysis was employed to analyse the data collected, which included students' photo essays and transcriptions of focus group discussions. Narrative analysis is a methodology in which "stories are used to describe human action" [52] (p. 5) and is grounded in the assumption that stories "constitute a fundamental form of human understanding, through which individuals make sense of themselves and of their lives" [53] (p. 22). Riessman [54] (p. 3) suggested that through the process of constructing a narrative, events perceived by a storyteller to be important may be "selected, organised, connected, and evaluated as meaningful for a particular audience". This approach required the researchers to focus on "particular actors, in particular social places, at particular social times" [55] (p. 428), and to construct their own narrative to detail the student participant's experiences in relation to the research question.

Polkinghorne [52] argued that these narratives are not meant to attempt to be objective or neutral depictions of the participant's life stories, rather their construction is influenced by the researcher's own experiences, views, and priorities. Indeed, they are influenced by the 'narrative terrain' of the research [56], i.e., the data produced through the dynamic and collaborative interactions between researchers and participants.

When adopting a narrative analysis approach, Sharp et al. [57] (p. 871) advised the use of direct quotes when developing key points to ensure participants' voices are present and to ensure "the narrative is grounded in data and authentic in tone". Overall, the analysis goal was to identify categories and themes common to the stories conveyed by the participating students through processes that included constant comparison [52].

9. Place as Complex Adaptive Assemblage: A Twofolds Analysis

Place as complex adaptive assemblage thinking [31–33] guided the narrative analysis process. The key to thinking about place as assemblage is the concept of twofolds. This suggests that places stabilise dwelling and being while promoting processes of becoming. Tensions operate at all places between a 'stabilised zone of order', which encompasses uniformity, familiarity, and fixed identities, and 'lines of movement', which promote the emergence of new identities. The 'stable' and the 'new' are not to be considered as conceptual oppositional binaries, rather they fold over each other in a complex mix of relations. Figure 3 represents Dovey's twofolds, which outline the characteristics of 'stabilised being' and 'processes of becoming'.

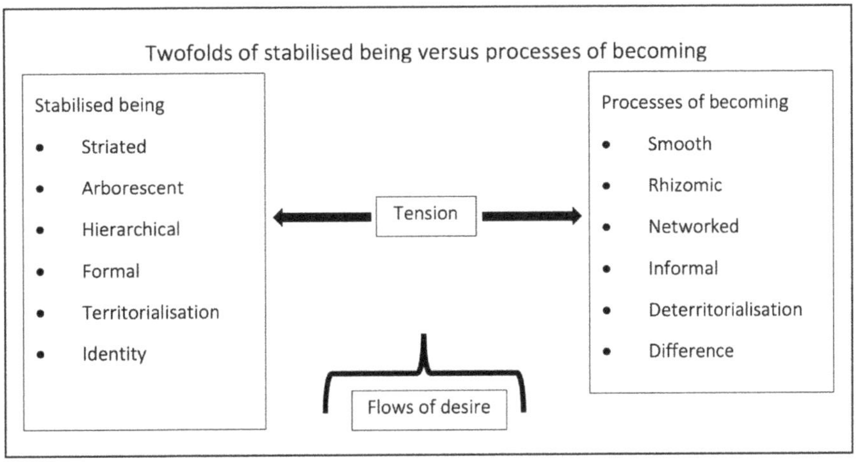

Figure 3. Twofolds of stabilised dwelling versus processes of becoming (adapted from Dovey [31–33].

Places offer stability by being fixed (striated), rooted and singular (hierarchical and arborescent), formal, and territorialised. Conversely, they promote becoming when there is a degree of 'slipperiness' in their lines of movement (smooth), and through being networked, rhizomic, informal, and deterritorialised (when erosion takes place). Informed by this line of thinking, being is not pre-given but is the outcome of becoming. As such, Dovey [31–33] suggests that it is through the ever-ongoing enfolding of the stable and the new that places and people become.

Places feature multiple parts and it is the relationships between these parts that stabilise being or promote becoming for different people. The relationships between a place's parts can be intense in an invigorating way or an oppressive way, imposing power over someone, or affording power to them. Alternately, there can be a resonance in the interaction amongst a place's parts, resulting in a coherence in a person's sense of place. Determining which relies upon a person's desire at that place. As such, the same parts, and the relationships between them, may result in varied senses of place. While commonalities may exist across different people's senses of place, everyone enters a place with their own desires, and it is these desires that engage with the relationships between a place's parts that informs a person's sense of place within a particular context [31–33].

To identify, evaluate, and compare the complexities inherent within students' photo essays according to place as complex adaptive assemblage thinking [31–33], an analytic frame was developed. This arranged the theory according to its six elements (power/desire, assemblage, territoriality/segmentarity, rhizome/tree, twofold, and sense). For each photograph and its accompanying text, images, words, or phrases were analysed using the frame to identify thematic clusters. This approach provided a robust method by which to analyse photographs and texts related to varied settings in each school with a degree of uniformity, thus enhancing reliability and replicability.

10. Biophilia and Adolescents' Sense of Place in Two Vertical Schools

Across the two case schools, the characteristics most commonly discussed by participants related to biophilia. All students identified biophilic elements when presenting their photo essays and were eager to share their personal responses to the elements they had identified. When they spoke of their relationships with biophilia, they described themselves, their emergence, settling down, and the formation of personal resilience. The types of relations students experienced with biophilia at each case school are outlined below.

10.1. City School

At City School, participants metaphorically reached out to the gardens adjacent to their school and commented that they felt "more open" because of them. They valued the "green" against the "grey" and expressed their love of lessons in the gardens. One student explained that the gardens are part of his school and that they make him feel "inspired". Another stated that the gardens represent "freedom" for her.

Such biophilic place relationships were mediated by the school's boundaries. The experience of the rhizomic, organic, supple, and soft biophilic elements of the gardens was valued by participants for its ameliorating effect on the city's striated, strict, rigid, and hard elements. Mediating this experience were multiple restrictions; some were due to the morphology of the school's building and its separation from the gardens by a busy city street, while others were due to the school's security practices.

The tension between rhizomic and striated, organic and strict, supple and soft—the "green" and "grey"—was experienced by each participant differently. Three participants were frustrated and felt their contact with the gardens was restricted and the boundaries imposed a "power over" them. Conversely, the remaining four students experienced a resonance between the gardens and the city, with one student particularly valuing the restrictions and perceived security offered by the school's building, promoting a "power to" her.

10.2. Drawn-Tight Boundaries and Restricting Access to Biophilia

Sam, Fiona, and Charlotte expressed a strong desire to connect to the gardens adjacent to their school. All three also expressed their desire for closer affective relationships with the gardens and commented on the limiting effect of the school's multiple boundaries, which they reported to exhort a power over them, restricting their agency and contact with nature.

The drawn-tight boundaries of the school's security practices were represented by Sam in his photograph of the gardens (Figure 4). Taken through a tiny hole, only millimetres in diameter, Sam pressed hard up against the balustrade of one of the school's outdoor terraces to photograph the gardens. He had a strong visual connection with the gardens but felt separated from them. He commented, "this photo feels as though the barricades enclosing the terrace separate the school". He added, "I guess it's like. . .um. . . that separation from the gardens". Sam felt restricted from the gardens and framed them by enclosing them in a ring of steel. Sam's photograph of student-only stairs (Figure 5) featured one of the school's security cameras prominently in the foreground. He commented that at school, "the cameras overlook everything", and that he felt "enclosed" and "isolated".

Fiona yearned for closer contact with the gardens. For her, they were boundless and seemed to go on forever. She commented that "freedom is out of reach", and she was "furious". Referring to the photograph in Figure 6, she commented, "there's the city in the background, City Gardens in the middle; freedom is out of reach; jealous of so many other schools with so much freedom and space to move around . . . we're confined to this small space". She directed her frustration at the school's "prison-like gates" that exerted "power over" her (see Figure 7). The gates were seen to dominate the school, were intimidating in scale, and removed agency from those entering and exiting the school. Fiona explained that "... it feels very prison-like; gates restrict us from being free; confused and furious. Why are we stuck in this building"?

Figure 4. View of City Gardens through a small hole from level four terrace (Sam).

Figure 5. Student-only stairwell (Sam).

While Charlotte was reassured by the school's security measures, she also agreed with Fiona, commenting that:

> There's obviously the idea of freedom because we can go out at lunchtime. But it also feels restricting because we have a security guard and stuff. There's a freedom of being in the city yet restricted by the gates. Isolated from the rest of the city ... there's obviously the idea of freedom, but it's kind of restricted at the same time and there isn't that much connection between us and the community.

Charlotte spoke of her lessons in the gardens and wished for more of them as she felt "more open", "inspired", and "optimistic" when there.

Figure 6. View from terrace (Fiona).

Figure 7. Security gates at entrance (Fiona).

10.3. Strong Visual Connection to Biophilia and a Resonance between the "Green" and the "Grey"

The remaining four participants described their affective relationships with the biophilic elements on offer in the gardens adjacent to the school. Lucas was proud of "his school's open fields" and "huge spaces" that were "never sad" and left him feeling "optimistic". He explained:

> My first photo shows a super extreme ... like hard ... view of the city. And it's like a really bad rainy day with the whole sky grey. Yeah, but I like the centre ... the image has City Gardens, which is always like super green. I've never seen it like sad ... except the trees in winter. What I like is that it conveys what it is like being schooled in the city, because it's with super contrast with all the other buildings.

The morphological boundaries and school security practices did not limit these four students in the same way that had reduced the agency of other students. All four valued the visual connections with the gardens, as afforded by the building's large windows. Charlotte loved lessons in the gardens, depicting her art lesson there (Figure 8). While Lucas observed a lesson through one of the windows and expressed that he felt the boundary of his school extended through the public gardens to the edge of the more distant city buildings. He described "the almost endless views to the buildings in the background ... almost making it feel like a border to our land" (Figure 9).

Figure 8. Charlotte's art class in City Gardens.

Figure 9. Lucas' observation of a class in City Gardens.

Both Lucas and Emma felt connected to the park, with Lucas valuing its "super green" in contrast to the "grey" city. Emma appreciated the "coming together" of the "man-made" elements and "nature" (Figure 10). She elaborated:

> Here, nature and the man-made environment have come together and we're lucky enough to have access to both . . . like an urban city and the materials from there and we've also got City Gardens as a natural park that we can go to hang out.

Ishan appreciated his strong visual connection with the gardens and enjoyed this experience with his friends from the eighth floor. During focus groups, he pushed back against the comments made by peers, asserting, "I do not feel totally disconnected to nature or isolated or trapped". Referring to the image in Figure 11, he commented, "It (the 'cloud shapes') look heavenly . . . they do look good . . . it's not like you're trapped in one building, right? Because you can go out and stuff". Ishan was "connected" to and appreciated the "nice-nice view" he and his friends had from the eighth floor at lunchtimes, where he felt "cosy" and "safe" (Figure 12). Ishan explained:

Figure 10. The "coming together of man-made and nature" (Emma).

Figure 11. "Heavenly" clouds (Ishan).

Figure 12. "Cosy" view (Ishan).

> There's like chairs and stuff which you would like. Sometimes at lunch you can go there and eat and you have this nice-nice view. It's really just about how we're not totally disconnected from nature because we can still . . . it's right next to us and we do go there like during lunch if you want to. It's pretty colourful and it's

> not totally isolated. You can see lots of City Gardens as you walk along Level Eight past the mirrors, which means you see lots of the gardens.

Freya went further than Ishan in expressing her appreciation of the school's security practices. They promoted "power to" her and she was "grateful" to have a visual connection to the gardens while being 'shielded from possible dangers' (Figure 13). She explained:

> I am really grateful that we get to have such a different view from all the other schools ... that would just have views of ovals and the ground level ... While we have views of the cityscape ... and City Gardens as well. The security measures taken when I enter the school make me feel safe and comfortable. I think the gates displayed at the front emphasise how much the school cares about the wellbeing and safety of its students, which brings me joy.

Figure 13. Visual connection with City Gardens from level seven (Freya).

10.4. Inner-Suburban High School

At Inner-Suburban High School, participants highly valued the incorporation of biophilic elements throughout their school and made personal connections with the interactions between the school's biophilic and manufactured elements. One student commented that without plants, school would be "boring". Another found the biophilic elements "beautiful" and "calming". Several students drew analogies between their emerging selves and the capacity for plants to grow, move, and change within a fixed setting.

The adolescents' biophilic place relationships were found to be mediated by the massive scale and volume of the school's buildings. A twofold relationship between the precision and permanence of the school's glass, concrete, and steel structures and the suppleness, growth, and change in the living plants was clearly apparent in the ways students discussed their experiences, also revealing a tension between the permanence afforded by manufactured parts, arranged in strict and precise ways, and the potential for something new and unknown afforded by plants and changes in the weather.

Desires and experiences of place varied across the five student participants. One found the scale of the facilities highly intimidating and limiting for herself and the plants that grow on and around the facilities, holding a "power over" her and them. Three found a resonance in the interplay between the manufactured and natural elements, with the building's scale seeming to not affect them. The fifth student suggested the potential for plants, especially trees, to have "power over" the buildings, despite their smaller scale.

10.5. Forceful Domination of Manufactured Parts over Natural Parts

Amelia felt her affective relationship with the biophilic elements at her school was constrained and exerted a forceful domination over her and the plants. She commented that what the school offers "is not really natural because it was just added there to make the school look better". She reported feeling frustrated by the presence of manufactured parts where "plants should be" (Figures 14 and 15). She explained:

Figure 14. Lawn, concrete garden edge, mulch, plants, and a grate (Amelia).

Figure 15. Concrete garden edge, plants, mulch and, red fire hydrant (Amelia).

> This is red, which is sort of like the opposite of green, if you think about it ... So, it's kind of like the red is the opposite of the greenery and it does not belong there. That is a bush out of the bush. So, the nature is confined to the space too and even where it is confined it still has things entering its space. It's basically the artificial things can go wherever it wants, but the nature can't go where it wants.

She resented the presence of a prominent drain (Figure 14), commenting that it "should not be where nature goes", and a red fire hydrant within a planting bed (Figure 15), "which is sort of the opposite of green". Just as the plants were displaced, so too was Amelia.

The school's entrance presented an intensity of order, rigidity, and scale (Figure 16) and she felt that agency had been removed from the plants and, by extension, herself. Amelia expressed her fear, intimidation, isolation, and loneliness at attending the vertical school. She said:

Figure 16. The entrance of Inner-Suburban High School (Amelia).

> When I first came to this school, I found it rather scary, then when I came for a second time, I was terrified again ... it was ... pretty scary because like, you know, it was loud, it was big. It's a big school, and it's very tall, which is rather different from other schools. So yeah, doesn't really have the feeling of a school.

Her photograph "Unbench" (Figure 17), showing a view from the underside of one of the school's benches, related to her experience of being dominated and overwhelmed at school.

Figure 17. View from under one of the school's benches (Amelia).

10.6. Resonance between Permanence and Precision of Manufactured Parts Versus Suppleness, Growth, and New Possibilities

Isabella, Camila, and Aisha expressed their affective relationship to the biophilic elements when they drew analogies between the plants and their identities. Camila appreciated the idea that just as plants were free to grow in "any direction", so was she. Aisha made a connection between plants' capacity for "messiness" within an ordered environment and her emerging self.

All three of these students found a resonance in the tension between the permanence and precision of the school's manufactured parts versus the plants' capacity to disrupt. Isabella found a degree of "calmness" in the "exactly parallel lines" of the built forms. Figure 18 shows red lines she added to her photograph, emphasising her appreciation of the school's uniformity. She explained:

Figure 18. Bench, rail, plants, and view to Brisbane. Red lines included by the participant (Isabella).

> Abra and I added the lines because it's got the horizontal lines. I'm kind of a perfectionist, so I like when they're exactly parallel, but the plants breaking it up is kind of nice. So it's a bit less harsh, because our school is quite harsh materials. Yeah, I do like lines. Why? I don't know. When they're parallel and they're perfect it makes me calm. It's weird because I like plants which are imperfect, but I like lines that are perfect. So I'm kind of contradicting myself there.

She went on to add that she likes the plants and the way they "break up the grey harsh materials" and that without them, school would be "boring". Isabella's description captured the resonance she experienced in the tension between her school's manufactured and biophilic elements.

Isabella, Camila, and Aisha also drew parallels between their growth and development and that of the school's plants. Isabella commented that just like her, the plants had grown considerably since she started school. Wrestling with her self-identity, perceptions of herself, and her direction in life, Camila made connections between the unpredictability of plant growth and her own development within a highly planned and structured environment when explaining her photograph, "Trees in the Sky" (see Figure 19):

> This photo is called "Trees in the Sky" because it's what it looked like . . . it's like a very specifically planned school . . . like all of the buildings in the background are very much architecturally planned, with the main plant in the front, that we have no way of planning where it's gonna go. Like, it's just gonna go wherever it wants to go. So, it's kind of showing the way that even though everything can seem very precise, it's always something you never know what's going to happen.

Similarly, Aisha commented about her propensity to be "neat" yet "messy", just like her school (see Figure 20). She said:

> So, the levels are kind of parallel and then there's stairs and green plants sticking out, and I guess that reflects me as a person. I seem neat sometimes, but then I have things sticking out and not perfect.

All three of these students found a personal resonance in the tension between the permanence, uniformity, and rigidity of the glass concrete and steel versus the suppleness and capacity for growth afforded by plants and how they promoted new possibilities. For them, their school's massive volume and scale seemed not to be a factor, due to the presence of biophilic elements.

Figure 19. 'Trees in the Sky': plants on the third-floor terrace with neighbouring buildings in the middle-ground and city as a backdrop (Camila).

Figure 20. "Messiness" of the Junior Hub (Aisha).

10.7. A "Foresty Vibe" Frames Manufactured Parts

Abra asserted that plants can frame, and perhaps even exert "power over" manufactured parts. While working on her photo essay, she realised the delight she found in one part of her school, especially relating to the way sunlight moves across the day (see Figure 21):

Figure 21. "Foresty vibe" (Abra).

> You can see the light crawling in from the right-hand side. This place never really had a meaning to me. Every morning, I would walk by it and like never pay attention to it. But when I got sent out to take photos, I really started paying attention to small spots of the school and realizing how beautiful they are. Although the school can be one of the places that most stress me out, moments that I capture like these really help me calm down.

Abra went on to describe the "shadowy places", "the green", the "shape of the tree", and the "blue sky" that help her feel calm when she is stressed:

> You can really see the green, and it's not a very bright photo. I love dark photos a lot, like shady places, and you can really see the contrast of the colours and the green leaves. It really kind of gives a foresty vibe. In this photo, the leaves kind of framed the building, I guess. And you can see all the contrast, much like all my other photos, with the blue sky and the railings and the green leaves.

The scale and volume of the school were not front-of-mind for Abra. For her, the school's biophilic parts dominated its manufactured ones. They afforded her a sense of calm and provided beauty in her everyday experience of school.

11. Discussion

The values students associated with their schools as places were found to be influenced by their desires. Some of the barriers at City School objected to by some students were embraced by others: restrictions and limitations for some were perceived to provide safety and security by others. Similarly, the scale and rigidity of Inner-Suburban High School's built form was rejected by one participant, yet its rhythmic and regular form were valued by other students: oppression for one, but consistency and precision for others. Common to the students' experiences of both settings were their schools' biophilic parts, which were valued explicitly or implicitly by all.

When describing their affective biophilic place relationships, students spoke of "freedom", "open fields", and "huge spaces" that "never feel sad". Students also expressed gratitude for strong visual connections to green spaces, and biophilic elements were identified as "beautiful", offering openness, inspiration, and optimism. Such values mirror those of Fromm's [7] biophilious person, as students expressed their desires for further growth, demonstrated their capacity to wonder and see something new, and indicated their love of the adventure of living.

Further, the biophilic parts of the two case schools were found to play a role in the ever-emerging processes of students' becoming through their capacity to promote resonances of tensions between the fixed and malleable: the existing and the new. The schools' becoming and the becoming of the students appeared to be influenced by the capacity of the biophilic parts to provoke new lines of movement. Participants' photo essays indicated that biophilic elements stimulated inventiveness and curiosity, prompted challenge, and even dissolved established boundaries in the minds of some. In the students' minds, each school's biophilic parts played a key role in promoting something new.

By contrast, the schools' manufactured parts, such as City School's boundaries and Inner-Suburban High School's scale and regular/rhythmic morphology, tended to stabilise dwelling. This provided uniformity, familiarity, and a sense of home in the experiences of several students. Yet, these same parts were also perceived as restrictive, limiting, and intimidating by others, as framed by different desires.

The biophilia associated with the two vertical schools tended to ameliorate the effects of their fixed manufactured parts, huge scales, and volumes—at least for most students. Biophilic elements proved in the minds of students to be formative parts of the place's resilience and that of their emerging adolescence. Dovey [33] argues that by considering place as a complex adaptive assemblage, that a regime or identity with certain characteristics may emerge, settle down, and become more or less resilient over time. Evidence of such developments was found in the students' photo essays, with students recognising established patterns of daily existence. For some, these created predictability and comfort, as they found resonance in the tensions and interplay between the school's permanent and precisely manufactured parts and the suppleness and capacity for growth, change, and new possibilities of plants and green spaces. While for others, such experiences caused stress and anxiety. Nevertheless, it was universally agreed that the biophilic contributions to these complex adaptive assemblages were inherently positive.

12. Limitations

By design, this research sought to understand the personal place relationships of a small number of adolescent students with their vertical schools. The research methodology enabled students to describe preferred and non-preferred parts of their school, as determined by their daily experiences and desires. While this resulted in deep reflections about a few of the schools' features, other features were not discussed. Additional students attending these schools may or may not have related to the same features in the same ways, or may have drawn attention to entirely different features of their school and associated place relationships. Nevertheless, the prominence of biophilic elements in the participating students' photo essays strongly indicated the importance of such features in their minds. It seems highly likely that if given the chance to contribute to the research, other students would have also featured similar biophilic place relationships, although perhaps with more varied interpretations. Follow-up studies may wish to develop larger data sets involving more students based on the insights gained through this small-scale qualitative exploratory study.

13. Conclusions

Given the rapid emergence of vertical schools in Australian cities, this research was timely as it provided insights into the place relationships adolescent students have with these urban environments, including how they relate to often limited outdoor spaces and reduced contact with natural elements when compared to more traditional 'low-rise' schools.

The findings demonstrated that the participating students not only valued relationships with natural elements, but actively sought biophilic experiences on a regular basis. The desire for more frequent and more significant interactions with natural materials and cycles was common across most students, indicating that the relatively limited biophilic experiences available to students in vertical schools is a challenge that should be addressed in future multi-level educational environments. While these findings may help sensitise architects, educators, and education authorities to opportunities for improved school planning and design, the focus here has been on students' place relationships and processes of becoming.

Adolescence is a 'messy business' and remaining in a stabilised state of being is not its objective. The malleability of plants and other biophilic parts, their capacity for change, movement, and growth in varied directions, along with their responsiveness to their circumstances, are all relatable for young people. Indeed, the students in this study strongly related their place experiences and the becoming of their identities to the biophilic elements in and around their schools, prompting these young people to reach out from their current states of being to something new. These processes of movement from the stable to the new aligned with the characteristics of Fromm's biophilious person [7] and highlighted the role of natural elements in the environment in the emergence of adolescents' diverse personal identities.

Based on the findings of this study, which indicate that biophilia in schools may influence students' place relationships and related identity formation, further research into adolescents' biophilic place relationships appears warranted. What effects are experienced and how these occur may be areas for further enquiry.

Author Contributions: Conceptualisation, methodology, analysis, investigation, original draft preparation, project administration and funding acquisition by T.B., guided by B.C.; review, editing and PhD supervision by B.C. All authors have read and agreed to the published version of the manuscript.

Funding: Funding gratefully received by Tim Baber as part of his PhD research includes: Australian Government Research Training Program Scholarship (fee offset and stipend); fieldwork grant and M.A. Bartlett Research Scholarship, both provided by University of Melbourne.

Institutional Review Board Statement: Research was conducted in accordance with the Declaration of Helsinki, with the approval of University of Melbourne's Office of Research Ethics and Integrity, reference number 2022-22460-25333-3, issued 18 January 2022.

Informed Consent Statement: Informed consent was obtained from all subjects (and their parents/careers) involved in this study.

Data Availability Statement: Data is not available for sharing due to ethical constraints.

Conflicts of Interest: The authors declare no conflict of interest.

References

1. Greater Sydney Commission. Central District Plan. 2020. Available online: https://www.planning.nsw.gov.au/sites/default/files/2024-04/central-city-district-plan.pdf (accessed on 3 March 2020).
2. Blandy, S. Is Your School Feeling the Pressure? The Population Experts. 2018. Available online: https://blog.id.com.au/2018/location-analysis/is-your-school-feeling-the-pressure/ (accessed on 2 May 2021).
3. Goss, P. Should You Worry about a School's Shortage? It Really Depends on Where you Live. The Conversation. 2016. Available online: https://theconversation.com/should-you-worry-about-a-schools-shortage-it-really-depends-on-where-you-live-53296 (accessed on 2 May 2021).
4. Carroli, L.; Willis, J.; Franz, J.; Barker, S.; Bone, N.; Brown, A.; Chisholm, N.; Gibson, A.; Gillett-Swan, J.; Harth, S.; et al. *Conversations about Vertical Schools in Australia*; Queensland University of Technology: Brisbane, QLD, Australia, 2022.
5. Taylor, H. High-Rise Schools. In *Urban Schools: Designing for High Density*; Taylor, H., Wright, S., Eds.; RIBA: London, UK, 2020; pp. 57–70.
6. Newton, C. Vertical schools on the rise in Australian cities. *Archit. Aust.* **2019**, *108*, 86.
7. Fromm, E. *The Anatomy of Human Destructiveness*; Macmillan: New York, NY, USA, 1973.
8. Wilson, E. *Biophilia*; Harvard University Press: Cambridge, MA, USA; London, UK, 1984. [CrossRef]
9. Kellert, S.R. Dimensions, elements, and attributes of biophilic design. In *Biophilic Design: The Theory, Science, and Practice of Bringing Buildings to Life*; Kellert, S.R., Heerwagen, J., Mador, M., Eds.; John Wiley & Sons: Hoboken, NJ, USA, 2013; pp. 3–19.
10. Kellert, S.; Wilson, E.O. (Eds.) *The Biophilia Hypothesis*; Island Press: Washington, DC, USA, 1993.
11. Ryan, C.O.; Browning, W.D.; Clancy, J.O.; Andrews, S.L.; Kallianpurkar, N.B. Biophilic design patterns: Emerging nature-based parameters for health and well-being in the built environment. *ArchNet-IJAR Int. J. Archit. Res.* **2014**, *8*, 62. [CrossRef]

12. Ryan, C.O.; Browning, W.D. Biophilic design. In *Sustainable Built Environments*; Springer: Berlin/Heidelberg, Germany, 2020; pp. 43–85.
13. Ghaziani, R.; Lemon, M.; Atmodiwirjo, P. Biophilic design patterns for primary schools. *Sustainability* **2021**, *13*, 12207. [CrossRef]
14. Watchman, M.; DeKay, M.; Demers, C.M.; Potvin, A. Design vocabulary and schemas for biophilic experiences in cold climate schools. *Archit. Sci. Rev.* **2022**, *65*, 101–119. [CrossRef]
15. Aminpour, F. Children's Preferences for Biophilic Design in Vertical Schools. *CTBUH J.* **2023**, 20–27.
16. Mohammed, I.; Onur, Z.; Çağnan, Ç. An Exploration of Biophilic Design Features within Preschool Interiors. *Sustainability* **2023**, *15*, 11913. [CrossRef]
17. Bagot, K.L.; Allen, F.C.L.; Toukhsati, S. Perceived restorativeness of children's school playground environments: Nature, playground features and play period experiences. *J. Environ. Psychol.* **2015**, *41*, 1–9. [CrossRef]
18. Collado, S.; Corraliza, J.A. Children's restorative experiences and self-reported environmental behaviors. *Environ. Behav.* **2015**, *47*, 38–56. [CrossRef]
19. Kelz, C.; Evans, G.W.; Röderer, K. The restorative effects of redesigning the schoolyard: A multi-methodological, quasi-experimental study in rural Austrian middle schools. *Environ. Behav.* **2015**, *47*, 119–139. [CrossRef]
20. Li, D.; Sullivan, W.C. Impact of views to school landscapes on recovery from stress and mental fatigue. *Landsc. Urban Plan.* **2016**, *148*, 149–158. [CrossRef]
21. Stenfors, C.U.D.; Van Hedgere, S.C.; Schertz, K.E.; Meyer, F.; Smith, K.E.; Norman, G.; Bourrier, S.C.; Enns, J.T.; Kardan, O.; Jonides, J.; et al. Positive effects of nature on cognitive performance across multiple experiments: Test order but not affect modulates the cognitive effects. *Front. Psychol.* **2019**, *10*, 1413.
22. Schertz, K.E.; Berman, M.G. Understanding Nature and Its Cognitive Benefits. *Curr. Dir. Psychol. Sci.* **2019**, *28*, 496–502. [CrossRef]
23. Wu, C.D.; McNeely, E.; Cedeño-Laurent, J.G.; Pan, W.C.; Adamkiewicz, G.; Dominici, F.; Spengler, J.D. Linking student performance in Massachusetts elementary schools with the "greenness" of school surroundings using remote sensing. *PLoS ONE* **2014**, *9*, e108548. [CrossRef]
24. Amicone, G.; Petruccelli, I.; De Dominicis, S.; Gherardini, A.; Costantino, V.; Perucchini, P.; Bonaiuto, M. Green breaks: The restorative effect of the school environment's green areas on children's cognitive performance. *Front. Psychol.* **2018**, *9*, 1579. [CrossRef]
25. Hodson, C.B.; Sander, H.A. Green urban landscapes and school-level academic performance. *Landsc. Urban Plan.* **2017**, *160*, 16–27. [CrossRef]
26. Li, D.; Chiang, Y.C.; Sang, H.; Sullivan, W.C. Beyond the school grounds: Links between density of tree cover in school surroundings and high school academic performance. *Urban For. Urban Green.* **2019**, *38*, 42–53. [CrossRef]
27. Erikson, E.H. *Identity: Youth and Crisis*; Norton: New York, NY, USA, 1968.
28. Branje, S.; De Moor, E.L.; Spitzer, J.; Becht, A.I. Dynamics of identity development in adolescence: A decade in review. *J. Res. Adolesc.* **2021**, *31*, 908–927. [CrossRef]
29. van Doeselaar, L.; Becht, A.I.; Klimstra, T.A.; Meeus, W.H.J. A review and integration of three key components of identity development: Distinctiveness, coherence, and continuity. *Eur. Psychol.* **2018**, *23*, 278–288. [CrossRef]
30. Bissell, D. Passenger Mobilities: Affective Atmospheres and the Sociality of Public Transport. *Environ. Plan. D Soc. Space* **2010**, *28*, 270–289. [CrossRef]
31. Dovey, K. *Becoming Places: Urbanism/Architecture/Identity/Power*; Routledge: London, UK, 2010.
32. Dovey, K. *Urban Design Thinking*; Bloomsbury: London, UK, 2016.
33. Dovey, K. Place as assemblage. In *The Routledge Handbook of Place*; Routledge: London, UK, 2020; pp. 21–31.
34. Heidegger, M. *Being and Time*; Harper Row: New York, NY, USA, 1962.
35. Relph, E. *Place and Placelessness*; Pion: London, UK, 1976; Volume 67, pp. 1–5.
36. Cresswell, T. *Place: An Introduction*; John Wiley & Sons: Hoboken, NJ, USA, 2014.
37. Malpas, J.E. *Place and Experience*, 2nd ed.; Cambridge University Press: Cambridge, UK, 2018.
38. Seamon, D. Place attachment and phenomenology: The dynamic complexity of place. In *Place Attachment*; Routledge: London, UK, 2020; pp. 29–44.
39. Massey, D. A global sense of place. In *The Cultural Geography Reader*; Routledge: London, UK, 2008; pp. 269–275.
40. Deleuze, G.; Guattari, F. *A Thousand Plateaus*; Athlone Press: London, UK, 1987.
41. McFarlane, C. The city as assemblage. *Environ. Plan. D* **2011**, *29*, 649–671. [CrossRef]
42. De Landa, M. *A New Philosophy of Society: Assemblage Theory and Social Complexity*; Continuum Press: London, UK, 2006.
43. Deleuze, G.; Parnet, C. *Dialogues II*; Columbia UP: New York, NY, USA, 1987; p. 69.
44. Holling, C.; Gunderson, L. Resilience and Adaptive Cycles. In *Panarchy*; Gunderson, L., Holling, C., Eds.; Island Press: Washington, DC, USA, 2002; pp. 25–62.
45. Johnson, S. *Emergence*; Penguin: London, UK, 2001.
46. Walker, B.; Salt, D. *Resilience Thinking*; Island Press: Washington, DC, USA, 2006.
47. Yin, R.K. *Case Study Research and Applications*; Sage: Thousand Oaks, CA, USA, 2018; Volume 6.
48. Appleton, J. *The Experience of Landscape*, Rev. ed.; Wiley: London, UK, 1996.
49. Google Maps. Map. 2024. Available online: https://www.google.com.au/maps/@-37.8111898,144.9548206,566m/data=!3m1!1e3?entry=ttu (accessed on 18 June 2024).

50. Google Maps. Map. 2024. Available online: https://www.google.com.au/maps/@-27.4959112,153.0266287,269m/data=!3m1!1e3?entry=ttu (accessed on 18 June 2024).
51. Lopez, E.; Eng, E.; Robinson, N.; Wang, C.C. Photovoice as a community-based participatory research method. *Methods Community-Based Particip. Res. Health* **2005**, *326*, 348.
52. Polkinghorne, D.E. Narrative configuration in qualitative analysis. *Int. J. Qual. Stud. Educ.* **1995**, *8*, 5–23. [CrossRef]
53. Ylijoki, O.-H. Master's thesis writing from a narrative approach. *Stud. High. Educ.* **2001**, *26*, 21–34. [CrossRef]
54. Riessman, C.K. *Narrative Methods for the Human Sciences*; Sage: Thousand Oaks, CA, USA, 2008.
55. Abbott, A. What do cases do? Some notes on activity in sociological analysis. In *What Is a Case? Exploring the Foundations of Social Inquiry*; Ragin, C.C., Becker, H.S., Eds.; Cambridge University Press: Cambridge, UK, 1992; pp. 53–82.
56. Holstein, J.A.; Gubrium, J.F. *The Active Interview*; Sage: Thousand Oaks, CA, USA, 1995.
57. Sharp, N.L.; Bye, R.A.; Cusick, A. Narrative Analysis. In *Handbook of Research Methods in Health Social Sciences*; Liamputtong, P., Ed.; Springer: Singapore, 2019; pp. 861–880.

architecture

MDPI

Article

Outcomes of Biophilic Design for Schools

William Browning [1,*] and Jim Determan [2,*]

[1] Terrapin Bright Green, 250 W 57th St., 17 FL, New York, NY 10107, USA
[2] Craig Gaulden Davis Architecture, 2002 Clipper Park Rd., Ste 201, Baltimore, MD 21211, USA
* Correspondence: bill@terrapinbg.com (W.B.); jdeterman@cgdarch.com (J.D.);
Tel.: +1-(646)-460-8400 (W.B.); +1-(410)-237-7115C (J.D.)

Abstract: Biophilia is the theory that humans are innately connected to nature. As a basis for forming built space, biophilic design has been proven to reduce stress, improve cognition and enhance mood—it makes us happier. In the last 20 years, research in learning spaces has shown an association between biophilic design and student mood, calmness and improved standardized test scores. In 2019, a group of architects, scientists and educators led an experiment involving 6th-grade Math students at the Green Street Academy, which found that student stress was significantly reduced and learning significantly improved in a classroom enriched with biophilic strategies. The architects applied these strategies to the design of Bethel Hanberry Elementary School, and after a year of occupancy, an independent assessment found positive perceptions of the biophilic design, fewer behavior referrals, better teacher retention, lower absenteeism and improved test scores. In both a controlled research experiment and real-world application, the design of learning space, using biophilic strategies, has a significant impact.

Keywords: biophilia; biophilic design; learning rate; school design; stress reduction

1. Introduction

Biophilia is the innate human connection to nature. The word was coined by the social psychologist Erich Fromm from the Greek words for life and love [1]. The most common definition comes from biologist Edward O. Wilson [2]. However, the history of designing for a connection with nature can be traced back millennia.

Research on how experiences of nature impact people looks at both psychological and physiological responses. It involves many fields of science, including environmental psychology, post-occupancy evaluation and neuroscience. Measurements can include heart rate, blood pressure, stress hormones, recovery time, fMRI imaging, eye tracking, cognitive performance and emotional responses.

The difference between experiencing a biophilic setting versus a non-biophilic setting can be measured through lower blood pressure, heart rate, and stress hormones, like cortisol, as well as through improvements in emotional state and short-term memory [3,4].

Being out in nature and even briefly viewing an image of a natural landscape can lead to experiencing a state called soft fascination. People are aware of their surroundings, but the brain is relaxed. This experience is part of Attention Restoration Theory [5]. When viewing nature, the prefrontal cortex of the brain quiets down and expends less energy; subsequently, when returning to focus on a task, there is improved cognitive capacity [6].

There are 15 different experiences of nature (with supporting scientific evidence of positive health impacts) that can be translated into experiences of the built environment. These experiences can be characterized into a pattern language for biophilic design [7]. These patterns fall into three broad categories: Nature in Space, Natural Analogues, and Nature of Space.

Nature in the Space refers to direct experiences of nature and natural processes in the built environment. These include views of the landscape, the presence of living plants, animals, water, sunlight, breezes, and the changing seasons.

Natural Analogues refers to indirect experiences of nature in the built environment. These include collinear and biomorphic forms, natural materials, and a level of complexity and order through materials or patterns, such as fractals.

Nature of the Space refers to experiences induced by certain three-dimensional characteristics of spaces. These include distant views through a space, spaces that provide refuge and retreat, spaces that compel exploration, spaces that have an element of risk and peril, and spaces that induce awe.

Different patterns support different outcomes. A pattern might support stress reduction, enhance cognitive performance, improve mood, or stimulate prosocial behavior. Some patterns support a single outcome, while others support multiple outcomes (Figure 1). A good approach to biophilic design is to create a narrative about the outcomes that you would like to support for the users of your space, and then determine which biophilic patterns best match those needs [7].

15 Patterns with Positive Health Outcomes

	Biophilic Pattern	Stress Reduction	Cognitive Performance	Emotion, Mood & Preference
Nature in the Space	Visual Connection with Nature	Heart rate, Blood pressure, Parasympathetic system activity	Mental engagement, Attentiveness	Attitude, Neurological rumination, Motivation, Future discounting
	Non-Visual Connection with Nature	Blood pressure, Stress hormones	Cognitive performance, Creativity	Perceived mental health, Tranquility, Pain management
	Non-Rhythmic Sensory Stimuli	Heart rate, Systolic blood pressure, Sympathetic nervous system		Dwell time, Behavioral attention and exploration
	Thermal & Airflow Variability	Comfort	Task performance, Productivity	Preceived temporal and spatial pleasure (alliesthesia)
	Presence of Water	Overall stress, Heart rate, Blood pressure	Cognitive performance, Creativity	Positive emotion, Tranquility
	Dynamic & Diffuse Light	Circadian system functioning, Visual comfort	Cognitive performance Behavioral performance	Attitude, Overall happiness
	Connection w/ Natural Systems	Overall health		Perception of environment
Natural Analogues	Biomorphic Forms & Patterns	Stress recovery	Learning outcomes	View preference
	Material Connection with Nature	Heart rate variability, Comfort, Calming, Blood pressure, Stress hormones	Task performance, Creativity	Material preference
	Complexity & Order	Perceptual and physiological stress responses	Environmental navigation, Learning outcomes, Mental relaxation	View preference
Nature of the Space	Prospect	Overall stress, Perceived safety, Comfort		Visual interest, Fatigue, Irritation, Boredom
	Refuge	Perceived safety		Visual preference
	Mystery			Pleasure response, Visual preference
	Risk/Peril			Pleasure response
	Awe	Stress related symptoms		Pro-social behavior, Attitude, Overall happiness

SOURCE: Browning & Ryan (2020). Nature Inside, A Biophilic Design Guide (Toolkit F)

Figure 1. Terrapin Bright Green, 15 patterns of biophilic design and associated outcomes [7].

2. Biophilia and Schools

Early research indicated that introducing daylight into classrooms improved academic performance [8,9]. Similar outcomes were seen in daylighting office spaces, although, over time, one researcher came to the conclusion that the view out the window might be more influential than just the effect of daylight. This led to an experiment in a daylit call center for a public utility company, in which workstations were realigned 11 degrees off perpendicular to the windows, to bring views to the outside within the peripheral vision of the workers. This resulted in a 6% improvement in call-handling capacity in the call center [10].

Classroom views to nature have been positively correlated to the likelihood of high school graduation and plans to attend college [11]. Students who do not graduate from high school typically earn about USD 10 k less per year than high school graduates and about USD 36 k less than college graduates [12].

The presence of nature in the schoolyard has also been studied. Access to nature, for example, tree cover in the schoolyard, has been shown to positively impact overall cognitive development, particularly students' working memory and attention [13].

Biophilic measures in the classroom have been found to improve student mood and calmness and to increase performance in standardized tests. Increasing student test scores by a quarter of a standard deviation would yield an average increase of 5.2% of discounted GDP above what is anticipated based on current achievement. The resulting increase would more than cover the entire U.S. expenditure on public education—approximately 4% of GDP [14].

An experiment in a girl's high school in London involved adding plantings to one classroom, a photo mural of woodlands in another, and no changes in a control classroom. Students in the classroom with plants demonstrated improved cognitive performance, and the students in the classroom with the photo mural indicated a higher level of emotional wellbeing [15].

3. Hypothesis and Methods

Our hypothesis is that simple low-cost biophilic design interventions could improve academic and other outcomes in schools. This paper summarizes the results of two studies. The first focused on a single classroom at the Green Street Academy in inner-city Baltimore, Maryland. The second study, building on the lessons of the first study, was an entirely new school, Bethel-Hanberry Elementary School, which replaced an existing school in Blythewood, South Carolina.

Our methods for measurement included comparisons of pre- and post-intervention academic outcomes, surveys and interviews of students and faculty. In the Baltimore study, biometric testing of stress recovery characteristics was conducted using heart rate variability (HRV) measurements. Since the Blythewood study involved an entire school, comparisons of absenteeism, behavior and teacher retention were also made.

4. Green Street Academy

The Green Street Academy is a locally funded public charter school in Baltimore, Maryland, in the United States. The school uses a STEAM (Science, Technology, Engineering, Arts, Mathematics) curriculum. The selected location for the school was in a shuttered City of Baltimore public school building in an underserved neighborhood. The 1925-era school was renovated and redesigned by architect Jim Determan of Craig Gaulden Davis Architects. Biophilic additions included indoor koi ponds, green houses and artwork of local ecosystems.

While the classrooms are not designed with specific biophilic elements, many of the classrooms are located on the long, east-facing rear façade and have a view out to grass and distant trees. The glare from the morning sun causes teachers to lower the window blinds, which are then rarely lifted.

Determan, along with partners from Morgan State University, Terrapin Bright Green, and the Salk Institute of Biological Studies, wanted to explore whether the use of low-cost biophilic interior elements would both lower the stress and improve learning outcomes in a classroom [16]. The classroom design interventions were focused on several biophilic design patterns, in particular Visual Connection to Nature, Biomorphic Forms & Patterns, Dynamic & Diffuse Light, and Complexity & Order.

To test Determan's hypothesis, a sixth-grade math classroom was renovated with carpet tiles, window shades, a wallpaper frieze and some waveform ceiling tiles, each expressing biomorphic forms or complex fractal patterns (Figure 2).

Figure 2. Craig Gaulden Davis Architects, Green Street Academy classroom redesign concept, Baltimore, Maryland, 2018–2019. Proposed biophilic classroom incorporating four simple design interventions.

The carpet tiles made by Interface featured a strongly collinear biomorphic pattern called Prairie Grass. The wallpaper frieze around the top of the classroom was jointly designed by artists from DesignTex/Steelcase and was an abstraction of palm leaves with biomorphic curves and collinear lines. These are examples of Biomorphic Forms & Patterns.

Many objects in nature have collinear striations or patterns that are broken into segments to form nested contours like fur, wood grain or grasses in a field. Studies with rhesus monkeys indicated that (within a given image) lines running in the same direction are processed by one set of neurons in the brain, whereas, with lines running in multiple directions, more effort is needed, by multiple sets of neurons, to process the image [17]. The brain will follow curvatures and contours [18] and even connect short segments of lines to discern a longer curving pattern [19]. These pattern conditions occur frequently in nature, and our brains, it could be argued, are predisposed to easily decipher them [20].

The venetian blinds were replaced with automated fabric blinds made by Mechoshade. The blinds had silkscreened patterns of tree branch shadows, which formed a pleasing statistical fractal pattern when the blinds were lowered. Fractals are layered self-repeating mathematical patterns. Exact fractals, which are the result of the same equation replicated at various scales, like embedded fractal gaskets or the trippy Mandelbrot sets, do not occur in nature. However, when those mathematical patterns have variations, their ubiquity in nature becomes quite evident, such as with snowflakes, fern leaves, waves on a beach, flames in a fireplace, the dappled light under trees. These are statistical fractals, so common that when we see these patterns, even in human-designed objects, it is easy for the brain to process the image and measurably lower our stress level [21,22] This effect is called fractal fluency [23]. This is an example of Complexity & Order.

The auto-controlled fabric window shades lifted, once the morning glare was off the windows, to a view of a newly planted outdoor garden. This supported both Dynamic & Diffuse Light and Visual Connection to Nature (Figure 3a).

For comparison, a seventh-grade math classroom along the same hall was chosen as a control space. The only change to that classroom was the addition of carpet tiles without any biophilic references so as to match the acoustic and haptic experiences of the experimental classroom (Figure 3b).

One class in the experimental classroom and another math class held at the same time in the control classroom were used for biometric testing. Heart rate variability (HRV) is a good indicator of the ability to recover from stress (high HRV is generally considered healthy). At the beginning and end of class, three times per week, student heart rate variability levels were measured using a finger-clip HRV monitor synced with a smartphone app. Due to the difficulty in getting permissions to conduct biometric testing with school children, there are only four months of HRV data. Students in the biophilic classroom demonstrated better stress recovery characteristics than students in the control classroom (Figure 4a).

(a)

Figure 3. *Cont.*

(b)

Figure 3. (**a**,**b**). Craig Gaulden Davis Architects, Green Street Academy classrooms, Baltimore, Maryland, 2018–2019. Comparative classroom study sites included two Math classrooms, (**a**) the biophilic classroom and (**b**) the control classroom. Blinds are drawn to show a comparative impact to the classroom environment.

During the 2018–2019 school year, the experiment tracked academic performance in a sixth-grade mathematics class. Test scores of current students were compared to those of students in the prior year in the same classroom with the same teacher and same curriculum. Students in the biophilic classroom attained better testing scores than the students in the prior years (Figure 4b).

Previously, the walls were covered with papers displaying formulas and notes, which some teachers will use to signal how much content is being conveyed in their classroom. Salk Institute neuroscientist Thomas Albright argues that covering the walls with so much material can result in an overstimulated environment.

Getting the teacher to agree to limit the amount of material taped onto the walls in the biophilic classroom was a challenge. At first, the teacher felt that would limit the students' retention of lessons, agreeing to put up sheets temporarily and then take them down. Over time, the teacher came to realize that the students calmed down in the room, she felt calmer in the room, and that other teachers would come into the room to take a break.

The view out of the window could have been a significant factor in the outcomes. However, for most of the experiment, there were no leaves on the trees. As seen in the HRV data, in April, when the trees blossomed, there was a spike in the response, which then diminished after the bloom. This is an indication that the biophilic design elements within the room were the significant factor in the outcomes [16].

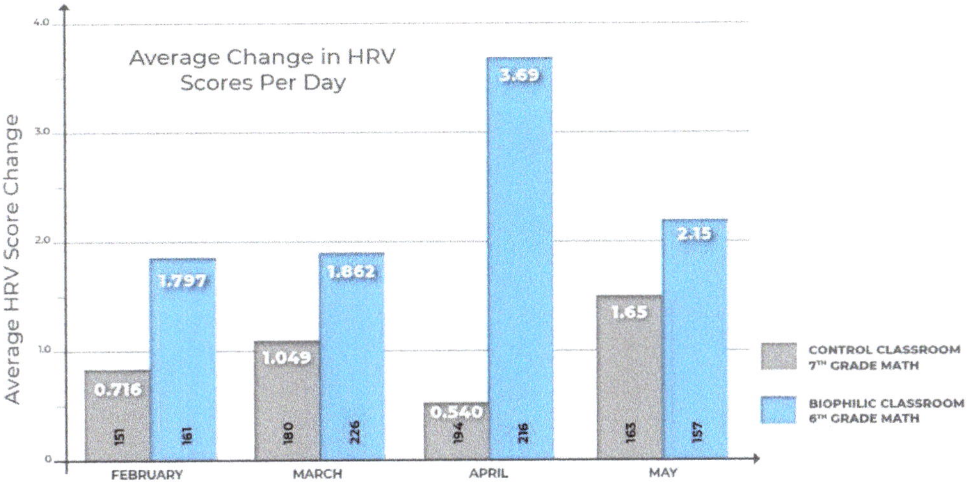

(a)

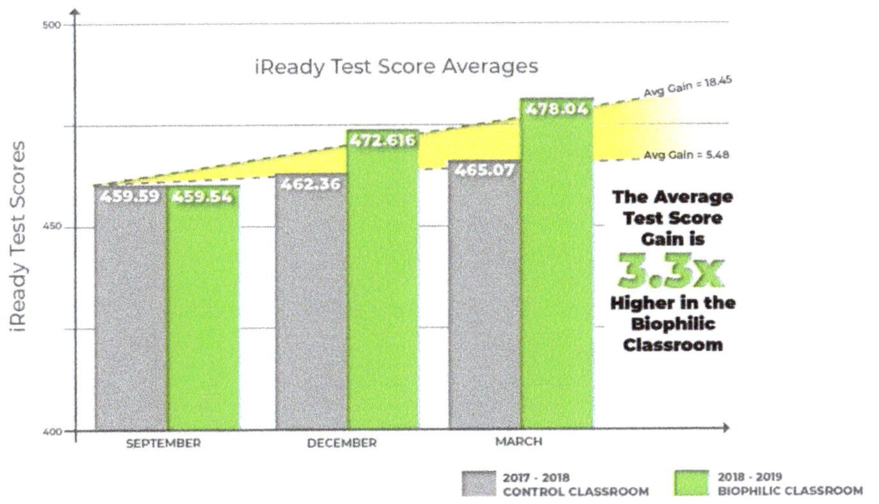

(b)

Figure 4. (**a**) Determan, et al. [16], Green Street Academy study results, 2018–2019. Heart rate variability (HRV). (**b**) Determan, et al. [16], Green Street Academy study results, 2018–2019. Test score differentials.

5. Bethel-Hanberry Elementary School

Design Application

Following the first study, the architects at Craig Gaulden Davis were excited to apply this new knowledge from the Green Street Academy Study to the design of Bethel-Hanberry Elementary School (BHES) in Blythewood, South Carolina. In this study, the entire population of students and teachers moved from an existing school into a new facility. What follows are illustrations of the biophilic applications.

Students enter a tall daylit space with wood panels above and biomorphic forms dancing along the walls. Prospect helps students feel safe as they can see through the perforated risers of a blue stair out to a courtyard (Figure 5).

Figure 5. Kris Decker/Firewater Photography, Bethel-Hanberry Elementary School Entrance Lobby.

Students walk through, what appears to be, the dappled light through a tree canopy. The shadow is cast by a ceramic frit pattern on the west-facing glass along the main corridor. This has a calming effect as students walk between classes in the afternoon. (Figure 6).

Long vistas create a feeling of safety, and window seats provide refuge along the main corridor (Figure 7).

Collaborative learning spaces are tall, open, and full of daylight, with views of nature. Transparency from the classroom to the collaboration space helps students feel safe knowing what is happening outside their classroom (Figure 8).

In the library, the grand scale, biomorphic forms and panoramic views of nature produce the 15th biophilic pattern, Awe. (Figure 9) Summer Allen, in *The Science of Awe*, [24] writes "Experiencing awe often puts people in a self-transcendent state where they focus less on themselves and feel more like a part of a larger whole". When it comes to psychological effects, studies have found that awe can "increase feelings of connectedness, increase critical thinking, increase positive mood". "Multiple studies have found evidence that experiencing awe makes people more kind and generous".

Figure 6. Kris Decker/Firewater Photography, Bethel-Hanberry Elementary School, Dynamic and Diffuse Daylight.

Figure 7. Kris Decker/Firewater Photography, Bethel-Hanberry Elementary School, Prospect + Refuge.

Figure 8. Kris Decker/Firewater Photography, Bethel-Hanberry Elementary School, Collaborative Learning Space.

Figure 9. Kris Decker/Firewater Photography, Bethel-Hanberry Elementary School, Awe.

6. Assessment

After the first year of occupancy at BHES, a team from the University of South Carlina Upstate conducted surveys and interviews to assess the impact of the biophilic design elements and used school metrics to assess student behavior and performance [25].

6.1. Perceptions

The percent of survey participants who agree that biophilic design strategies have made a positive contribution to the learning environment is as follows.

Parents 95%

Students 82%
Teachers 93%
Administrators 100%

6.2. Absenteeism

Out of 20 elementary schools in the district, BHE is tied for having the greatest reduction in chronic absenteeism (missing 10% or more of the school year), reducing from 17.3% (last year in the old school) to 12.3% (this year in the new school). The National Center for Education Statistics reports that the national average of chronic absenteeism is 17% [26], and the SC Department of Education reported an average of 24.73% for the 2021–2022 school year.

6.3. Behavior

The Assistant Principal reports that there are fewer disciplinary problems this year as compared to last year. "There's been a decrease in the overall number of referrals and the intensity of the infractions. I think the sense of calm that's promoted throughout the building makes a difference. And I think it manifests in the behaviors that we do and do not see".

6.4. Teacher Retention

BHES experienced among the highest increases in teacher retention of the 24 elementary schools in the district. Teacher retention increased from 83.7% to 91.5%. Districtwide teacher retention was 81.9%.

6.5. Learning Outcomes

MAP (Measures of Academic Progress) testing measures student growth in Math and Language Arts/Reading and compares student growth from one year to the next. BHES MAP scores in Math indicate that growth exceeded projected growth in every grade. The Language Arts/Reading scores exceeded the projected growth for grades 2–4 and met the projected growth in grade 5.

Principal Holland describes the academic performance improvement the first year in the new school as "fantastic and highly motivating!". It is amazing that we met all and exceeded seven of eight growth projections. This is a great accomplishment, which we celebrate and are energized to continue. This is remarkable in the context of a 2023 NAEP report that scores have declined in every course nationally.

While the designers do not that claim biophilic design strategies are solely responsible for the student success described in this assessment, given the research and survey results, they believe biophilic design made a powerful contribution.

7. Discussion

Biophilic design is sometimes viewed as requiring living interventions like green walls and live animals. Green walls and live plants can both be biophilic strategies. There is evidence that live plants in a classroom can enhance cognitive performance [15]. There is evidence that animals in the classroom can have a positive effect on children [27]. Neither of the studies in this paper used these living interventions. Both relied on passive measures that were part of the space; different outcomes might be possible with other design strategies.

In both studies, academic outcomes improved, and additional benefits were measured. Inevitably, improvements in performance in a changed or new facility are attributed to the Hawthorne Effect, which is typically summarized as changes in the environment signal concern for the users, thereby, lead to gains in performance, and, therefore, the physical environment is less important than how people are managed. While datasets in changed environments will show a spike in performance in the first blush of time, after a few months, the newness wears off. Yearlong datasets overcome that effect [28], and other investigations of productivity have come to different conclusions than the Hawthorne work [29,30]. Additionally, investigation of the original Hawthorne studies indicate that they were deeply flawed and do not support their widely quoted conclusions [31,32].

While the biophilic design interventions in both studies used surfaces with biomorphic forms and statistical fractals and other design strategies, it is unclear which elements were most effective in supporting a biophilic response in the spaces. Research using gaze attention tracking tied with physiological measurements in a virtual office space indicates that this combination of measurement technologies can provide indications of which biophilic design interventions are most effective in a specific space [33]. Future research using gaze attention tracking tied with physiological measurements in real spaces could provide an indication of which interventions are most effective.

8. Conclusions

From research experiments to real-world application, we see the benefits of biophilic design in learning spaces. If minimal design interventions in the classroom can help make students happier, healthier and improve learning, and if they can improve teacher retention, reduce absenteeism and curtail bad behavior, why would we not do this in all schools? Every effort to help improve a young person's capacity to learn and enhance their social-emotional wellness will pay dividends to them as individuals and for us as a society.

Schools at the turn of the 20th century were daylit and naturally ventilated by operable windows. This typically allowed for views of trees, clouds and other outside activity. In the 1960s, the U.S. education system perpetuated the belief that views were 'distractions' and that the attention of children should be focused on activities within the classroom. Henceforth, the construction of windowless or transom-window classrooms became pervasive in practice, particularly among temporary modular structures, many of which are still in use decades later. Bringing access to daylight, views to nature and the use of biophilic elements in the classroom clearly have a measurable benefit for the learning environment [34].

David Orr, an emeritus professor of Environmental Studies at Oberlin College, says that buildings are 'crystalized pedagogy'; they inherently tell us about the belief systems that led to their design [35]. Biophilic design can be leveraged as a philosophy of education and design that helps both support improved academic performance and a greater connectedness between buildings and nature.

Author Contributions: Writing—original draft preparation, W.B. and J.D.; writing—review and editing, W.B.; visualization, J.D.; All authors have read and agreed to the published version of the manuscript.

Funding: The research for the Green Street Academy Study was partially funded by a 2017 UpJohn Research Grant from the American Institute of Architects.

Institutional Review Board Statement: This paper is a summary of prior research papers. The Green Street Academy study was conducted with the approval the of Baltimore City Public Schools Board of Education and the Institutional Review Board of Morgan State University. The Bethel-Hanberry School study was conducted with the approval the of Richmond County School District, and the University of South Carlina Upstate.

Informed Consent Statement: Informed consent was obtained from all subjects involved in the study.

Data Availability Statement: Details of the Green Street Academy study can be found in the paper Determan, J.; Akers, M.A.; Albright, T.; Browning, B.; Martin-Dunlop, C.; Archibald, P.; Caruolo, V. *Impact of Biophilic Learning Spaces on Student Success*; American Institute of Architects, Building Research Information Knowledgebase: Washington, DC, USA, 2019. Which can be accessed at https://www.brikbase.org/content/impact-biophilic-learning-spaces-student-success (accessed on 28 April 2024). Details of the Bethel-Hanberry Elementary School post occupancy study have not been published, and were obtained through O'Connor, J.W.; O'Connor, C.B. Elevating Learning Environments through Biophilic and Student Centered Designs. *Preprint* 12 March 2024.

Conflicts of Interest: William Browning is an employee of Terrapin Bright Green, and Jim Determan is an employee of Craig Gaulden Davis Architects are the authors. The authors declare that the research was conducted in the absence of any commercial or financial relationships that could be construed as a potential conflict of interest. The funders had no role in the design of the study, in the collection, analyses or interpretation of data; in the writing of the manuscript, or in the decision to publish the results.

References

1. Fromm, E. *The Heart of Man: Its Genius for Good and Evil*; Harper & Row: New York, NY, USA, 1964.
2. Wilson, E.O. *Biophilia, The Human Bond with Other Species*; Harvard University Press: Cambridge, MA, USA, 1984.
3. Yin, J.; Zhu, S.; MacNaughton, P.; Allen, J.G.; Spengler, J.D. Physiological and cognitive performance of exposure to biophilic indoor environment. *Build. Environ.* **2018**, *132*, 255–262. [CrossRef]
4. Park, B.J.; Tsunetsugu, Y.; Kasetani, T.; Morikawa, T.; Kagawa, T.; Miyazaki, Y. Physiological Effects of Forest Recreation in a Young Conifer Forest in Hinokage Town, Japan. *Silva Fenn.* **2009**, *43*, 291–301. [CrossRef]
5. Kaplan, S. The restorative benefits of nature: Toward an integrative framework. *J. Environ. Psychol.* **1995**, *15*, 169–182. [CrossRef]
6. Lee, K.; Williams, J.H.; Sargent, L.D.; Williams, N.S.G.; Johnson, K.A. 40-second green roof views sustain attention: The role of micro-breaks in attention restoration. *J. Environ. Psychol.* **2015**, *42*, 182–189. [CrossRef]
7. Browning, W.; Ryan, C. *Nature Inside, a Biophilic Design Guide*; Royal Institute of British Architects: London, UK, 2020.
8. Nicklas, M.H.; Bailey, G.B. Analysis of the performance of students in daylit schools. In Proceedings of the National Passive Solar Conference, Asheville, NC, USA, 13–18 April 1996; pp. 132–137.
9. Heschong, L. Day lighting and student performance. *ASHRAE J.* **2002**, *44*, 65–67.
10. Heschong, L. *Windows and Offices: A Study of Office Worker Performance and the Indoor Environment*; Heschong Mahone Group, California Energy Commission, Pacific Gas and Electric Company: Fair Oaks, CA, USA, 2003.
11. Matsuoka, R.H. Student performance and high school landscapes: Examining the links. *Landsc. Urban Plan.* **2010**, *97*, 273–282. [CrossRef]
12. Bureau of Labor Statistics. Current Population Survey. Median Weekly Earnings (in Dollars) and Unemployment Rate (in Percent) by Educational Attainment. 2017. Available online: https://www.bls.gov/careeroutlook/2018/data-on-display/education-pays.htm (accessed on 28 April 2024).
13. Dadvand, P.; Nieuwenhuijsen, M.J.; Esnaola, M.; Forns, J.; Basagaña, X.; Alvarez-Pedrerol, M.; Rivas, I.; López-Vicentea, M.; De Castro Pascual, M.; Su, J.; et al. Green spaces and cognitive development in primary schoolchildren. *Proc. Natl. Acad. Sci. USA* **2015**, *112*, 7937–7942. [CrossRef] [PubMed]
14. Hanushek, E.A.; Ruhose, J.; Woessmann, L. Economic Gains from Educational Reform by US States. *J. Hum. Cap.* **2017**, *11*, 447–486. [CrossRef]
15. Bowman, C.; Bowman, R.; Bowman, D. *The Biophilic Classroom Study*; Putney High School Girls' Day School Trust: London, UK, 2019. Available online: https://www.gdst.net/publications/the-biophilic-classroom (accessed on 28 April 2024).
16. Determan, J.; Akers, M.A.; Albright, T.; Browning, B.; Martin-Dunlop, C.; Archibald, P.; Caruolo, V. *Impact of Biophilic Learning Spaces on Student Success*; American Institute of Architects, Building Research Information Knowledgebase: Washington, DC, USA, 2019.
17. Gilbert, C.D. in Intermediate-level visual processing and visual primitives (Chapter 27). In *Principles of Neural Science*, 5th ed.; Kandel, E.R., Schwartz, J.H., Jessell, T.M., Siegelbaum, S.A., Hudspeth, A.J., Mack, S., Eds.; McGraw Hill: New York, NY, USA, 2014.
18. Li, W.; Gilbert, C.D. Global contour saliency and local collinear interactions. *J. Neurophysiol.* **2002**, *88*, 2846–2856. [CrossRef] [PubMed]
19. Li, W.; Piech, V.; Gilbert, C.D. Learning to link visual contours. *Neuron* **2008**, *57*, 442–451. [CrossRef]
20. Albright, T.D.; Stoner, G.R. Contextual influences on visual processing. *Annu. Rev. Neurosci.* **2002**, *25*, 339–379. [CrossRef] [PubMed]
21. Hägerhäll, C.M.; Laike, T.; Taylor, R.S.; Küller, M.; Küller, R.; Martin, T.P. Investigations of human EEG response to viewing fractal patterns. *Perception* **2008**, *37*, 1488–1494. [CrossRef] [PubMed]
22. Hägerhäll, C.M.; Laike, T.; Kuller, M.; Marcheschi, E.; Boydston, C.R.; Taylor, R.S. Human physiological benefits of viewing nature: EEG responses to exact and statistical fractal patterns. *Nonlinear Dyn. Psychol. Life Sci.* **2015**, *19*, 1–12.
23. Abboushi, B.; Elzeyadi, I.; Taylor, R.S.; Sereno, M. Fractals in architecture: The visual interest, preference and mood response to projected fractal light patterns in interior spaces. *J. Environ. Psychol.* **2019**, *61*, 57–70. [CrossRef]
24. Allen, S. *The Science of Awe*; Greater Good Science Center, University of California Berkeley: Berkeley, CA, USA, 2018.
25. O'Connor, J.W.; O'Connor, C.B. Elevating Learning Environments through Biophilic and Student Centered Designs. *Preprint* 12 March 2024. Available online: https://cgdarch.com/wp-content/uploads/2024/04/BHE-Assessment_3.14.pdf (accessed on 28 April 2024).
26. US Department of Education. Chronic Absenteeism in the Nation's Schools. 2019. Available online: https://www2.ed.gov/datastory/chronicabsenteeism.html (accessed on 28 April 2024).

27. Gee, N.C.; Griffin, J.A.; McCardle, P. Human-Animal Interaction Research in School Settings: Current Knowledge and Future Directions. *AERA Open* **2017**, *3*, 2332858417724346. [CrossRef]
28. Romm, J.; Browning, W.D. *Greening the Building and the Bottom Line*; Rocky Mountain Institute: Snowmass, CO, USA, 1994.
29. Brill, M.; Margulis, S.T.; Konar, E. *Using Office Design to Increase Productivity, Volume I*; Workplace Design and Productivity, Inc.: Buffalo, NY, USA, 1984; pp. 224–225.
30. Dickson, W.J.; Roethlisberger, F.J. *Counseling an Organization: A Sequel to the Hawthorne Researches*; Harvard University Press: Boston, MA, USA, 1986.
31. Parsons, H.M. What Happened at Hawthorne? *Science* **1974**, *183*, 922–932. [CrossRef] [PubMed]
32. Parsons, H.M. What Caused The Hawthorne Effect? A Scientific Detective Story. *Adm. Soc.* **1978**, *10*, 259–283.
33. Yin, J.; Yuan, J.; Arfaei, N.; Catalano, P.J.; Allen, J.G.; Spengler, J.D. Effects of biophilic indoor environment on stress and anxiety recovery: A between-subjects experiment in virtual reality. *Environ. Int.* **2020**, *136*, 105427. [CrossRef] [PubMed]
34. Ryan, C.O.; Browning, W.D.; Walker, D.B. *The Economics of Biophilia: Why Designing with Nature in Mind Makes Financial Sense*, 2nd ed.; Terrapin Bright Green, LLC.: New York, NY, USA, 2023. Available online: http://www.terrapinbg.com/report/eob-2 (accessed on 8 July 2024).
35. Orr, D. *Earth in Mind*; Island Press: Washington, DC, USA, 1994; p. 113.

architecture

MDPI

Review

The 'Nature' of Vertical School Design—An Evolving Concept

Alan J. Duffy

Architectus, Sydney, NSW 2000, Australia; alan.duffy@architectus.com.au

Abstract: Successful urban school design includes green space to counterpoint the built form in cities, where parks and reserves are well frequented. Further integration of landscape and buildings is an aspect of urban development that could improve how architecture is experienced by the wider community. Above all, evidence shows that it enhances the health and wellbeing of inhabitants. By providing green space in buildings, nature can be accessed more directly by its occupants and allow connection with nature to occur more easily. Integrating nature with architecture can improve a building's self-regulation, energy consumption, and overall performance. Architecture that integrates nature can have a distinctive appearance and character. The co-existence of bricks and mortar with plants and vegetation is one example of integration, whereas the use of natural materials such as timber as part of the building fabric can create distinctive architecture. It is this individuality that can provide a sense of identity to local communities. Access to the outdoors in urban settings is a critical requirement for successful urban school design. This paper focuses on the architectural practise of designing biophilic schools and illustrates how optimising playground opportunities can provide the highly sought-after connection between architecture and nature. Connecting classrooms and pedagogy to the outside environment during the design phases of projects can create unique responses to a place, enhancing the learning experience in environments where architecture and nature can be informed by emerging biophilic evidence. This study strives to develop a strategy where educational clients can be convinced to actively embrace a biophilic school approach. It also seeks to convince architects to adopt a biophilic approach to school design across design studios using the emerging evidence based on biophilia and biomimicry.

Keywords: biophilic; design; architecture; nature; education; learning; students; wellness; health; wellbeing

1. Introduction—Why Nature in School Design?

The research question framing this paper asks, 'what is the evidence to support the inclusion of nature in schools, how can this academically rigorous evidence be better distilled to inform school architectural design practises and their clients, and how can this movement be scaled up in schools' architectural design and educator professions?'

In particular, the dual aim of this study is to provide a translational research pathway for evidence-based academic biophilic/biomimicry design research to find its way into architectural design studios and into educators' own pedagogical practises.

This article is organised into five sections. Section 1 explores why we should include nature in school design and how the nature-positive movement can form the basis of this proposition. This then informs the potential impacts on the built environment and underpins biophilic design principles. Section 2 suggests a methodology to suit a design practitioner's evidence-based approach informed by nature-positive biophilic design processes. It introduces an established biomimicry design process model which commences with the identification of the spatial requirements of functional teaching and learning. This step is supported by the establishment of the evidence base for the use of biophilia and its impact on learning. These concepts are then applied to the biophilic design of vertical schools using exemplars of prior learning landscapes followed by the evaluation of the ongoing evolution of the design. Section 3 elaborates on the idea of positive nature informing design strategies. This includes the use of typologies using nature as a design narrative. Section 4 presents a discussion which considers pedagogy and play space; the use of daylight, fresh air, and views of nature; and the final building expression, all of which can have an effect on students' learning outcomes and their health and wellbeing. Section 5 concludes with a strong argument for a return to nature in the design of schools for health and wellbeing.

1.1. The Nature-Positive Movement

The association between human health, wellbeing, and the natural environment is well established through empirical research [1]. In the context of architecture, the implications for best practises are clear—championing more nature-based design through the inclusion of biophilia is important for living, working, and learning environments. This idea is being advocated through the 'nature-positive' movement [2].

Biophilia is described as the innate tendency of humans to focus on life and life-like processes, meaning we have an instinctive desire to socialise ourselves in the natural environment and connect with other living organisms such as trees, plants, and animals. The theory was first publicised in 1964 by Erich Fromm [3], a US social psychologist, and popularised in 1984 by Edward Wilson [4], a biologist at Harvard University. Both emphasised that connection with nature leads to a better response in terms of human functioning, health, and wellbeing.

Population movement in Australia is exponential and mostly involves people moving to metropolitan areas, where competition for space is producing more high-rise buildings, including schools [5]. Swollen urbanisation with town planning densification policies and the locations of working environments in response to hybrid digital integration are some determinants influencing time spent in and out of natural environments [6].

Despite the positive relationship between time spent in nature and improved community health, people spend a significant amount of time (>80%) in buildings, separated from the benefits that nature provides. They also spend increasing amounts of time in an inactive mode while being 'glued' to screens.

1.2. The Impact of the Built Environment

The topic concerning the health and wellbeing of urban dwellers, children in schools, and employees in commercial towers is gaining more attention. Many studies have shown that connecting with natural environments or 'green space', including urban green space and agricultural space, promotes good health and is highly effective in reducing symptoms of stress and mental fatigue [7].

Excessive time periods spent in buildings can lead to Sick Building Syndrome (a group of factors that can negatively affect health in different ways), which can be costly for the community and economy due to sickness, employment absence, and reduced productivity [8]. For example, air-conditioned buildings are known to have a higher prevalence of symptomatic employees compared to naturally ventilated buildings.

Biophilic design aims to mediate some of the negative consequences that built environments can have on human health (including psychological and physiological health) through facilitating modifications to building design in line with human needs. Health aspects related to access to fresh air, natural daylight, and views of nature have been shown to improve respiratory health, reduce mental fatigue and feelings of stress, and restore attention capacity [9]. Building design can improve by increasing understanding about biophilia and its benefits to building occupants.

1.3. Biophilic Design Principles, Nature, and Architecture

Biophilic design principles can include but are not limited to the following: (i) access to natural daylight, (ii) access to fresh air, (iii) direct access to living nature, (iv) views of green spaces, (v) the use of natural materials, (vi) designing for [acoustics] sound, (vii) the presence of water, and (viii) proximity to restorative (and natural) outdoor spaces. The application of these and other biophilic principles can help improve the design of new schools and support educational communities [10].

Successful urban design includes green space to counterpoint the built form in cities, where parks and reserves are well frequented. Further integration of landscape and buildings is an aspect of urban development that could improve how architecture is experienced by the wider community. By providing green space in buildings, nature can be accessed more directly by its occupants and allow connection with nature to occur more easily [1].

Integrating nature with architecture can improve a building's self-regulation, energy consumption, and overall performance [11]. Architecture that integrates nature can have a distinctive appearance and character. The co-existence of bricks and mortar with plants and vegetation is one example of integration, whereas the use of natural materials such as timber as part of the building fabric can create distinctive architecture. The Macquarie University Incubator building for business start-ups in Sydney is a fine example of the latter. It is this individuality that can provide a sense of identity to local communities [12].

2. Methodology—A Nature-Positive Design Methodology for Educational Campuses

Architectural designers operate in a hybrid fashion, seeking evidence to support their design decisions while aligning with the client's value proposition for each particular project. In some cases, they use rigorous academic evidence-based research, and in other cases, they might supplement those sources with built precedents from published architectural journals (not necessarily scholarly). They are also obliged to follow policies and procedures which are mandated by school authorities at various levels.

The application of biophilia in the design process requires the designer to offer a robust argument to curate and shape a new way of thinking in terms of classrooms engaging with nature. To this end, using an established and tested biophilic model is a useful way to engage with both designers and educational clients.

2.1. *Applying Biophilic Principles in Design through the Concept of Biomimicry*

Biomimicry is a design process which seeks to understand natural health and well-being drivers and how these can be embedded in the architectural design process. The Biomimicry Institute developed a five-step biophilic design process called the biomimicry design spiral, as illustrated in Figure 1 [13,14].

This concept was proposed to simplify the connection between existing and emerging rich, extensive, and deeply rigorous academic research so that it can inform architectural practitioners in a more simplified yet evidence-based way. The model ensures that designers use an ecosystem approach in their design decisions, with all of them being based on rigorous evidence.

The key steps include the following:

(a) Distill—what do you want your design to do, and why?
(b) Translate—what are the functions and how can nature habitat support this?
(c) Discover—biophilic taxonomies, benchmarking, literature reviews, case studies, evidence.
(d) Emulate—consider Terrapin's 14 elements and case studies [15].
(e) Evaluate—is the solution organically biophilic?

The following section is a suggested framework for a biomimicry school design process using this model.

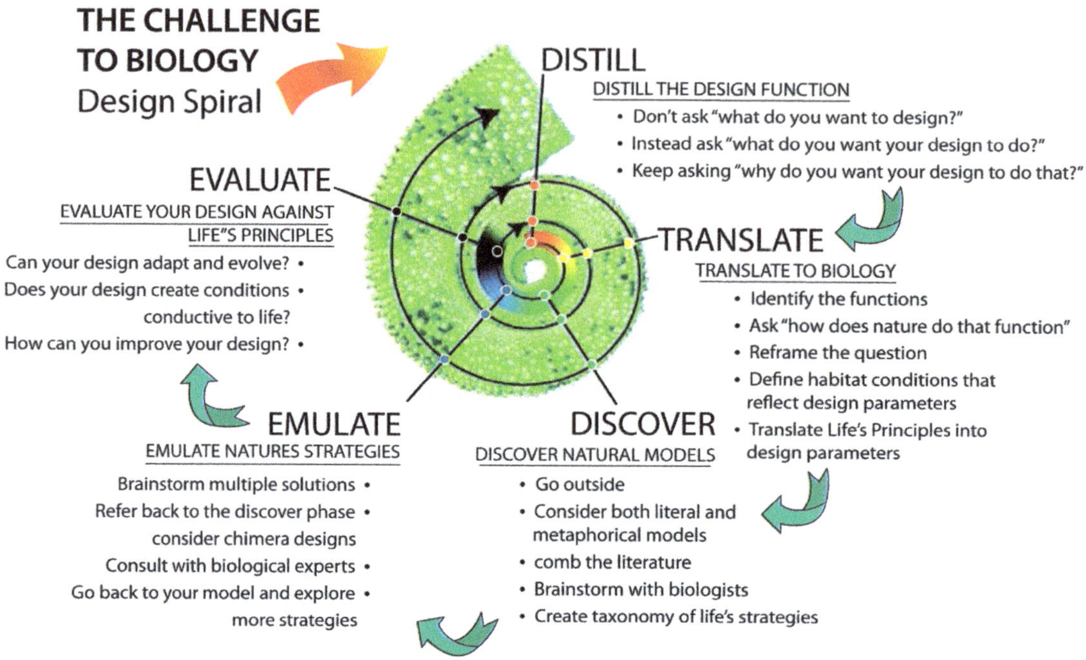

Figure 1. Biomimicry design spiral (source: McGregor, 2014).

2.2. Distilling Functional Requirements

We are now designing schools for the most technologically savvy demographic to date—the so-called generation alpha. These students were born after the iPhone was released [16]. Their cultural perspective of smartphones, tablets, technology, and media is re-shaping traditional classroom layouts and settings. In addition, emerging innovation in artificial intelligence, augmented reality, virtual reality, mixed reality, and robotics is curating and moulding educational experiences and learning practises. The integration of these tools in classrooms helps motivate and engage students in new ways that are quickly replacing [old school] didactic/instructional methods.

Despite advancements in digital integration and new technology, studies indicate that by the time children reach adolescent years, screen time increases to, on average, between seven and ten hours per day [17]. Australian research has shown that excessive use of technology among students is linked to issues such as increased isolation, depression, anxiety, and attention problems along with the largely sedentary nature of the digital world [18]. Biophilic design in education may counteract some of these issues by reconnecting students to nature and alleviating some of the burden associated with increased technology use.

Architecture that reflects nature can contribute to students' learning experiences by framing the space in which education occurs. The sizes, shapes, and aspects of rooms and the selection of building materials and interior fabrics are some features that can improve learning spaces for students. Balancing the presence of technology with nature is an important consideration for integrating student engagement and coherence in classroom settings. Modest interventions, such as the option to open or close a window, can provide the ability to adjust the immediate environmental conditions, supporting health and wellbeing in classrooms.

2.3. Translate—Student Engagement and Learning in Nature

This is carried out through the process of translating biomimicry into the design process, as noted earlier. Cultivating and attaining student engagement is a common goal in almost all schools and education settings. Positive psychology models, such as PERMA (Positive Emotion; Engagement; Relationships; Meaning; and Accomplishment), used by the New South Wales Department of Education [19], aim to cultivate attributes that foster student engagement, confidence, and improved learning experiences that lead to higher levels of student success. Nature fundamentally supports many aspects that encourage these educational goals. The school playground is a space that successfully dovetails biophilia and education theory in the outdoors, where a sense of connection and collaborative learning is accessible through social interaction and direct exposure to nature.

Access to natural settings is important for all students at all levels of education. For example, local schools in Sydney often identify with unique natural elements specific to their locality, such as eucalyptus or fig trees or resident school ground animals such as rabbits. Natural features can help form a sense of identity for school students and create a common binding interest for school communities—where nature itself becomes a classroom/learning environment. The diversity of natural habitats across NSW provides abundant opportunities to integrate biophilia into many schools through design strategies that consider the integration of the built form and natural surroundings more coherently.

Associations between indoor/outdoor spaces and the influence on learning outcomes is an evolving research area. New studies have shown that there are correlations between more contact with nature and improved academic performance. As an example, students who are exposed to more natural daylight in classrooms have been shown to perform better academically compared to students with less access to natural light [20]. Biophilic design is important in education, particularly in areas related to the following:

- Social and emotional learning: Experiences of natural environments have been shown to support greater social and emotional restoration and reduce tension and anxiety, confusion, anger, as well as fatigue compared to urban environments [21], helping with students' attitudes, behaviour, and academic performance.
- Memory and attention: One hour in nature improves memory performance and attention by 20%. Research that involved participants who walked in urban city streets and botanical gardens (nature) showed that short-term memory improved by 20% among participants who walked through gardens compared to those who walked in urban streets, which showed no improvement [22].
- Fatigue: The Attention Restoration Theory (ART) suggests that concentration fades when performing cognitive tasks using directed attention over sustained time periods [23]. ART suggests that nature provides the specific environmental stimuli to facilitate recovery from issues such as attention fatigue.

2.4. Discover—Introducing Biophilia into Vertical School Designs

The idea of a healthy school and campus has origins in the open-air school movement in the early 1900s in Europe [24]. Educationalist Walter Spickendorff and paediatrician Prof. Dr. Bernhard Bendix in collaboration with Berlin's Schools Inspector Hermann designed an 'open-air therapy' environment for students with pre-tuberculosis called the Waldschule für kränkliche Kinder (forest school for sickly children) in 1904. Around the same time in the late 19th C, the idea that 'form follows function' was raised by Louis Sullivan [25]. A modernist approach that ensures the function of internal space takes precedence over the built form, ornamentation, or aesthetic of the building. This philosophy has influenced architecture for over a century, inspiring buildings that are designed from the inside out. This has recently evolved into a more nuanced and elemental approach that considers the functional affordances on campus [26].

Tall buildings remain centric in contemporary urban settings and now also exist in educational environments including schools [27]. As the learning landscape evolves, improving student-centred learning remains in focus. Architecture can help to achieve many aspects of this through the application of biophilia—bringing the outside in, diffusing boundaries between classrooms and the outdoors, and creating opportunities for students to connect with nature.

Vertical schools can exploit height by providing unique views of nature and new outdoor experiences above the natural ground level. Courtyard settings with access to green space, trees, and communal gathering is possible whether it is at the natural ground level or elevated above street level by providing outdoor settings with the presence of living nature and, preferably, tree canopy cover. These elevated courtyard spaces are not a replacement for open school playgrounds; however, they can provide much needed relief from artificial enclosures used during school learning hours and offer a more immediate connection with nature within bigger buildings.

This can help support cognitive function and ability for students and provide a restorative function during the day's learning activities. The potential for this architectural approach to enhance school design is exciting because it can develop a new student experience that supports health and learning as well as provide variance to the building's character. It is also likely to foster greater student engagement in learning.

Figure 2 shows an artist's impression of an urban school with multiple levels, where typical arrangements of classrooms can connect directly to outdoor spaces over several levels.

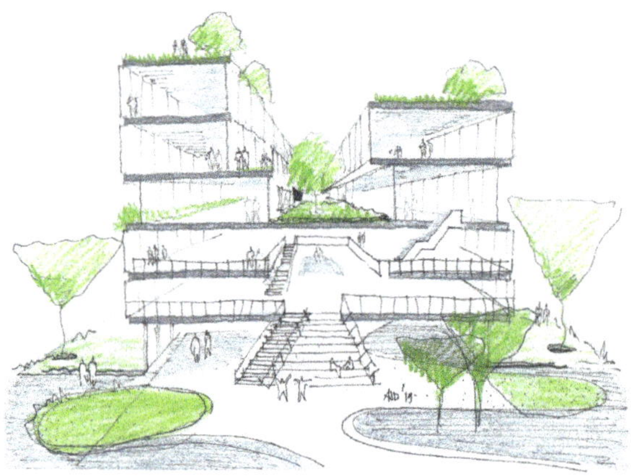

Figure 2. Urban school with elevated outdoor space (source: author, 2019).

This concept relates to schools in dense urban areas, where student cohorts can enjoy large contiguous learning zones by blurring boundaries between inside and outside. Playful connecting elements, such as stairs, lifts, and playground equipment, can provide character and identity to each floor level and the overall school building, where learning activities and movement can be expressed through the architecture.

2.5. Emulate—Learning Landscapes

Nature has always been a consistent source of inspiration for human endeavours, propelling our understanding and interaction with the world. Scientists, inventors, and designers have continuously looked to nature for guidance in solving problems [28]. An outdoor learning space that is designed in harmony with nature can support learning and wellness through a curated approach to landscape design. With Science, Technology, Engineering, and Mathematics (STEM), a growing focus area for school curricula worldwide, school buildings themselves become teaching and learning tools.

The life systems of the building itself can teach students about architecture, engineering, and environmentally responsive design. From water consumption to electricity generation, from service reticulation to structural engineering, and from natural ventilation to high-performing building façades, the building itself becomes part of the learning experience. Technology can provide a bridge between the natural and the human-made by capturing and displaying information regarding how the building functions and operates, its impact on the environment around it, and the building's performance as a learning tool.

Many students may remember a teacher who influenced their thinking, education journeys, and sometimes even their career choices. Great teachers can spark curiosity in young learners and give them the confidence to believe in themselves to pursue their dreams and goals. Future generations of scientists, technology wizards, talented engineers, or gifted mathematicians can be wholly supported in a well-designed learning landscape inspired by great teachers and in harmony with nature. The natural environment as the primary engineer showcases complex biodiversity in balanced ecosystems. Similarly, architecture and nature can contribute to the student experience when integrated, whereby human-made and natural engineering co-exist in harmony with one another.

The use of space in between general learning environments and traditional classrooms is widely known to evoke knowledge sharing and peer-to-peer learning through chance encounters and social interactions, where learning through conversation can occur. Learning in corridors and informal settings has been shown to improve learning ability and build relationships in education settings. When the space in between extends to the outdoors, opportunities for learning and wellbeing increase [29]. Outdoor classrooms in the right climatic conditions can be very successful; however, local environmental conditions do not always favour outdoor learning. Thresholds between indoor and outdoor spaces can offer additional diversity with informal settings, where learning can occur in breezeways, circulation zones, and covered outdoor learning areas particularly suited to inclement weather.

2.6. Evaluate—Evolving Design Concepts

School design continues to evolve as a response to changing urban conditions, including the emergence of precincts, hubs, and high-density living in cities. Connecting the classroom to the outdoors becomes a key challenge for urban school projects, where site constraints limit access to outdoor space due to increasing land value and decreasing land available for development. Inventive design solutions to meet playground provisions for students and provide access to landscape settings for teachers requires careful planning.

Design strategies that capitalise on accessing urban outdoor space often involves the consideration of elevated playgrounds and other complex design solutions, including associated connecting voids, staircases, and safety enclosures (refer to Figure 6). Designing areas with direct access to outdoor spaces that are safe and protected, with natural materials and access to daylight, direct sunlight, and natural breezes, can inspire creative thinking and bespoke design solutions. The study of specific site conditions and local weather patterns of a particular place also becomes an important part of the design process.

Effective outdoor learning spaces can be tested for climatic conditions that are expected to be experienced during the various seasons. The integration of elevated playgrounds to meet various student learning activities can yield unique learning settings, including terraced outdoor spaces, covered outdoor learning areas (COLAs), and interconnecting stairs and light wells. In protecting and enclosing the perimeter of elevated play spaces, the building façade design process for urban school buildings presents creative opportunities to respond to the multipurpose activities occurring on different floor levels.

3. Nature-Positive Design Strategies

3.1. Typologies

Design decisions that consider nature can lead to positive outcomes in the built environment. Four effective strategies that can promote qualities of improvement in school design and help support student-centred learning as well as health and wellbeing include the following:

1. The connected façade: This involves providing openable doors and windows in all buildings. Having a direct connection between the indoor classroom and a green outdoor space can have a significant impact on wellbeing—see Figure 3.
2. The window seat/nook: This is an intimate setting to encourage peer-to-peer and self-directed learning for students. Access to daylight views of nature in a comfortable

window seat is an effective example of a well-designed space that can support positive outcomes in behaviour and learning. This also enhances the sense of territorial privacy while keeping an observational distance from supervisors—see Figure 4.

3. The courtyard: As a well-established architectural typology, the courtyard design creates a safe communal outdoor space that is both protected by and connected to indoor learning spaces (this can be on the ground level or at an upper level of a high-rise building). Attributes of biophilic design are well supported within the courtyard model, where access to outdoors, fresh air, daylight, and green space and the presence of water are all easily accommodated in a private setting that is suitable for several uses, including group gatherings, outdoor classrooms, and individual contemplation and refuge. In particular, the cloister and/or veranda and colonnades or arcades also provide a transition from inside to outside, with this transitional space being ideal for nature-oriented activities in inclement weather.
4. Natural materials: The use of natural materials in architecture is successful because of their tactile nature and richness in texture and colour, particularly when it comes to wood. The use of timber for furniture, joinery, and interior surfaces including floors, walls, and ceilings can improve the interior quality of schools and bring nature right to the heart of the learning environment. New building techniques being used which dovetail the latest technology with natural materials, such as cross-laminated timber (CLT), are now being considered for their benefits in prefabrication and precision, their speed in construction, and importantly, their natural material, timber. The use of exposed timber can greatly improve the interior aesthetics of learning facilities, adding qualities of colour, texture, odour, and acoustics and improving the overall spatial experience for occupation.

Figure 3. The connected façade and courtyard space. (Source: author, 2019).

Figure 4. The window nook (source: author, 2019).

These validated approaches are commonly found in vernacular architecture all over the world, from Japanese Zen gardens to the cloisters of Europe. Applying similar design principles on a larger scale, specifically involving taller and more complex school buildings, is important in the context of promoting healthier environments for students in city locations and high-rise developments.

3.2. Nature as a Design Narrative

Universities have traditionally enjoyed large, landscaped grounds for their campuses; however, land for new development is increasingly scarce. This can result in larger, taller buildings being used for adult education. Health and wellbeing are at the helm of many research institutions and affiliated universities. Therefore, designing buildings with access to nature to improve student and employee wellbeing has become more important. The idea of research and learning in comfortable settings and buildings that respond to location, cultural heritage, and campus history can enhance staff, student, and researcher experience (see Figure 5) [30]. When considering the design approach for a building, the process can involve research into the site's history, the cultural context, and the existing building character surrounding the proposed site.

Story-telling and design narrative play important roles in this process. The design intent can be communicated, encouraging imagination to envision the proposed building with the fullest understanding of the factors influencing the design concept. Nature can inspire this narrative, where a site's history can be explored in terms of its geology, ecology, cultural heritage, and architectural lineage in relation to the urban surroundings.

In this context, the proposed architecture and natural environment formulate a synergistic relationship. The boundaries between architecture and nature can elicit building designs that are grounded in their context and connected to their natural environment, both in narrative and built forms.

Figure 5. Flinders University's Health and Medical Research Building, Adelaide, South Australia. (Source: author, 2024).

Consciously designing with culture and heritage as a priority often leads to a fuller expression of a building's unique character. The completion of Flinders University's Health and Medical Research Building is an example of designing for place while being informed by consultation and thoughtful conversations with local indigenous elders, including stories about country [31]. Notwithstanding the building's predominant research focus, the architecture creates moments of connection to the natural context, where land, water, and sky inform the architecture and physical form of the building.

The built form is shaped in layers, reflecting the natural terracing of the Adelaide foothills and sedimentary geological processes depicting timelines in the earth. The building is layered and terraced to enjoy expansive vistas towards the seaboard and the vast Southern Ocean horizon. Connections from the internal spaces and outdoor landscapes, features of dedicated winter gardens, and a large cascading public plaza anchor this building in its unique context.

3.3. *Implications of Climate and Climate Change on Designing for Health and Wellbeing*

Biophilic design can depend upon climatic conditions. Tropical climates (such as Northern Australia) are more suitable for year-round outdoor access and support continuous plant growth throughout the year, providing more opportunities for planting integration around and on top of building structures. Allowing the teaching and learning experience to occur in environments that are generously landscaped in regional climatic and specific site conditions can provide unique experiences for building occupants, where they can inhabit natural outdoor spaces in an individually distinctive architecture. Figure 6 illustrates a biophilic approach to a vertical campus building within a tropical climate region, where stacked terrains form the building itself.

Figure 6. An illustration of a biophilic building concept for a tropical climate (source: Architectus, 2022).

Schools are striving towards greater sophisticated buildings to support learning and education outcomes, particularly secondary schools that are increasing STEM-based opportunities that require technology-enabled buildings to support students' readiness for tertiary education. Learning spaces that resemble university or workplace settings may enhance the student experience and familiarity with these environments prior to the transition from secondary school to university.

The transition from senior school to tertiary education environments is an important development stage due to the highly technical and digital characteristics associated with STEM [29]. Student health and wellbeing is an important variable in this progression. Biophilic design which encourages health-oriented natural environments has been shown to fortify student resilience and encourage performance [32,33]. The implementation of this design methodology could support future expectations, attitudes, and success in tertiary education environments.

4. Discussion

Realising the benefits of nature and placing biophilic principles into practise can take many forms for a school building depending on student profiles, curriculum, site location, and context. With particular emphasis on urban schools, pedagogy and play spaces ought to be considered together because it is in these environments that the balance between indoor and outdoor spaces is most compromised.

4.1. Pedagogy and Play Spaces

With team teaching practises and collaborative learning approaches being pursued in New South Wales state schools in Australia, classroom configurations and adjacencies with shared learning spaces can influence how the overall school is designed. School Infrastructure New South Wales (SINSW) implemented a learning hub model where four classrooms work as a collaborative learning unit, and they are connected via shared learning spaces centred between the four classrooms (refer to Figure 7).

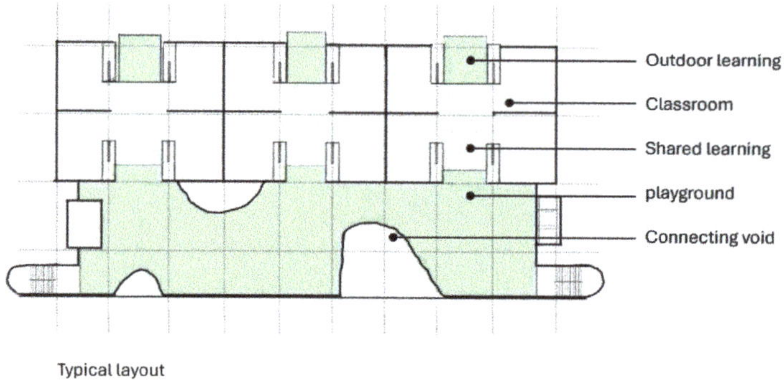

Figure 7. Urban school plan layout diagram. (Source: author, 2022).

For a primary school in an urban setting, this generalised approach can lend itself suitable for a multi-storey vertical building, with playgrounds on ground level and again on the rooftop, to maximise site coverage. Many urban sites are heavily constrained regarding space to build on and occupy. Outdoor spaces beyond the building footprint and landscaped zones are often limited due to this constraining factor. Figure 6 displays a potential approach to address this issue. By designing elevated outdoor spaces and integrating them into school floor plates, student cohorts can occupy dedicated floor levels, each with their own outdoor play space that is directly accessible from the indoor classroom environment. This provides a learning landscape on each level, protected from the urban setting, while encouraging connection and collaboration for the students both indoors and outdoors.

4.2. *Daylight, Fresh Air, and Views of Nature*

With play spaces allocated to each level, providing ample daylight, filtered sunlight, and fresh air becomes an important aspect to achieve. The placement of light wells or voids on elevated floor plates is a design strategy that can bring playful geometry, engaging connections between floors and allowing for lighting conditions to be changed throughout the day. This interaction with sunlight and the passing of the day can encourage better awareness of the natural environment.

Student health can also be fortified through this awareness and direct access to daylight, nurturing a healthy circadian rhythm and overall wellbeing for students. Having opportunities to view green spaces with ecological diversity is another key consideration in early design phases for school buildings, ensuring that the benefits of surrounding landscapes can be harnessed and used to their best potential for the school and local community, evoking better school environments with healthier classrooms for students of all ages.

4.3. *Building Expression*

Vertical schools in dense urban environments can be unique, individual buildings expressing the activities within and providing a civic presence for communities [33]. With a particular emphasis on playgrounds over multiple levels, the traditional building façade can become a distinctive architectural feature. The play space becomes the building façade when viewed from street level, activating the streetscape with student movement and interaction, where play is clearly on display at several intervals throughout the day. This animated building edge can be a playful, artistic, and colourful feature to explore in urban school design. Rather than having walls and windows looking into classrooms, the building presents as a multi-storey playground, as illustrated in Figures 7–9.

Figure 8. An urban school where the play space becomes the building façade. (Source: author, 2022.)

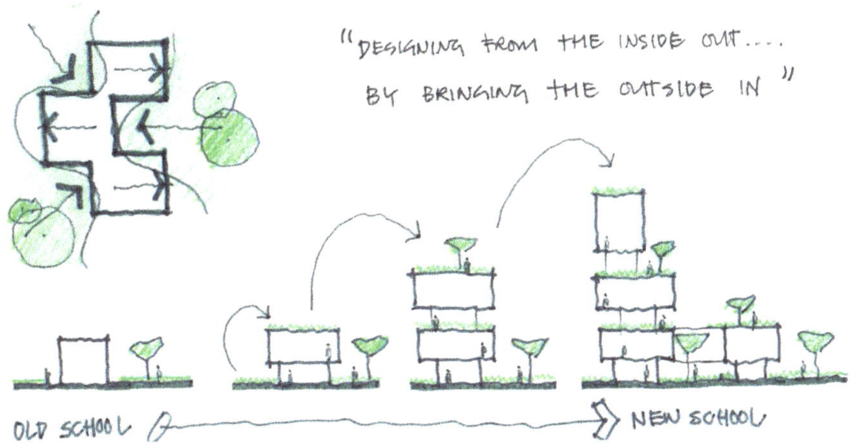

Figure 9. Designing from the inside out by bringing the outside in—biophilic urban schools. (Source: author, 2019.)

Many of these ideas will be developed over time as designers seek to evolve their designs in coming years [34]. Indeed, as further research is conducted with the aim of understanding the relationship between school building design and student and staff health and wellbeing, these concepts will be refined and adapted over time from an increasing evidence base in support of such outcomes [35].

5. Conclusions—Back to Nature

In aspiring to assist educators in improving student health and wellbeing and ultimately improving learning outcomes through biophilic design, school designers are seeking ways to mimic nature in their built projects. A design approach that considers the overall user experience, including that of students, educators, staff, and the community, can yield positive results through appropriate exposure to nature. Aspects such as daylight, fresh air, and greenery can affect how students hear, breathe, see, feel, and move and can influence their ability to absorb, understand, and apply learning in the school environment. Nature is an asset to embrace as it can bring aspects of balance and harmony to the classroom.

Designing from the inside out by bringing the outside in, as illustrated in Figure 9, can unify human-made and natural environments.

Applying biophilic design principles through a biomimicry lens has the potential to transform the nature of new school design by integrating function, form, and the student experience in schools, helping to restore architecture that promotes health and wellbeing in educational environments, particularly in dense urban locations. That said, biophilic design is yet to make a significant impact on the minds of senior figures of authority in educational systems globally. The mantra of form and function does not seem to include nature and its impact on health and wellbeing. This is a major barrier which this article seeks to address through supporting designers in creating strategies to convince educational authorities to adopt this approach.

The benefit of a design-led biophilic/biomimicry campaign to achieve this engagement is that these academic concepts—largely textual and numerical in form—can be visualised in concept designs. These designs, when built, can be used as case studies, as illustrated by Browning and Determan, to convince authorities to embrace this approach. This article is essentially about science communication, an area which continues to struggle to translate complex scientific research into applications that can be made useful for society in practise. Future 'research' might focus more deliberatively on the communication of the science of biophilic/biomimicry design practises and applications to educators and design practitioners alike.

Funding: This research received no external funding.

Institutional Review Board Statement: Not applicable.

Informed Consent Statement: Not applicable.

Data Availability Statement: All data used during the study are presented in the paper.

Conflicts of Interest: The author is a Principal and shareholder at Architectus. The author declares that the research was conducted in the absence of any commercial or financial relationships that could be construed as a potential conflict of interest. There was no funding provided for this article.

References

1. Beatley, T. *Handbook of Biophilic City Planning & Design*; Island Press: London, UK, 2016.
2. Oxford Hub. Nature Positive: A Definition. 2024. Available online: https://naturepositive.web.ox.ac.uk/nature-positive-oxford (accessed on 7 July 2024).
3. Fromm, E. *The Heart of Man: Its Genius for Good and Evil*; Open Road Media: New York, NY, USA, 2023.
4. Wilson, E. *Biophilia*; Harvard University Press: Cambridge, MA, USA, 1984.
5. Australian Government. Intergenerational Report. 2010. Available online: https://treasury.gov.au/sites/default/files/2019-03/IGR_2010.pdf (accessed on 23 April 2024).
6. Candido, C.; Durakovic, I.; Marzban, S. *Routledge Handbook of High-Performance Workplaces*; Routledge Handbooks: London, UK, 2024.
7. Grahn, P.; Stigsdotter, U.A. Landscape planning and stress. *Urban For. Urban Green.* **2003**, *2*, 1–18. [CrossRef]
8. Kinman, G.; Clements, A. The Role of Demographic and Psychosocial Factors in Predicting SBS Symptoms in Workplaces. In *Sick Building Syndrome*; Kinman, G., Ed.; Springer: Berlin/Heidelberg, Germany, 2011; pp. 393–404.
9. Berto, R. Exposure to restorative environments helps restore attentional capacity. *J. Environ. Psychol.* **2005**, *25*, 249–259. [CrossRef]
10. Kellert, S. Building for Life: Designing and Understanding the Human-Nature Connection. *Renew. Resour. J.* **2005**, *24*, 8.
11. Browning, W.; Kallianpurkar, N.; Ryan, C.; Labruto, L. *The Economics of Biophilia Why Designing with Nature in Mind Makes Financial Sense*; Terrapin Bright Green: New York, NY, USA, 2014.
12. Macquarie University Incubator. 2024. Available online: https://macquarie.architectus.com.au/project/incubator-macquarie-university/ (accessed on 7 July 2024).
13. McGregor, S. Transdisciplinarity and biomimicry. *Transdiscipl. J. Eng. Sci.* **2014**, *4*, 57–65. [CrossRef]
14. Pawlyn, M. *Biomimicry in Architecture*; RIBA Publishing: London, UK, 2016.
15. Browning, W.; Ryan, C.; Clancy, J. *14 Patterns of Biophilic Design: Improving Health & Well-Being in the Built Environment*; Terrapin Bright Green: New York, NY, USA, 2014.
16. Rideout, V.; Foehr, U.; Roberts, D. *Generation M^2: Media in the Lives of 8- to 18-Year-Olds*; Henry J. Kaiser Family Foundation: San Francisco, CA, USA, 2010.

17. Carson, V.; Hunter, S.; Kuzik, N.; Gray, C.E.; Poitras, V.J.; Chaput, J.-P.; Saunders, T.J.; Katzmarzyk, P.T.; Okely, A.D.; Gorber, S.C.; et al. Systematic review of sedentary behaviour and health indicators in school-aged children and youth: An update. *Appl. Physiol. Nutr. Metab.* **2016**, *41*, S240–S265. [CrossRef]
18. Thomas, G.; Bennie, J.A.; De Cocker, K.; Ireland, M.; Biddle, S. Screen-based behaviours in Australian adolescents: Longitudinal trends from a 4-year follow-up study. *Prev. Med.* **2020**, *141*, 106258. [CrossRef] [PubMed]
19. NSW Department of Education. Positive Psychology Models Such as PERMA (Positive Emotion; Engagement; Relationships; Meaning; and Accomplishment). 2024. Available online: https://hurlstone-h.schools.nsw.gov.au/learning-at-our-school/perma.html (accessed on 7 July 2024).
20. Dahlan, A.; Eissa, M. The impact of daylighting in classrooms on students' performance. *Int. J. Soft Comput. Eng. (IJSCE)* **2015**, *4*, 2231–2307.
21. Tanner, C. Explaining relationships among student outcomes and the school's physical environment. *J. Adv. Acad.* **2008**, *19*, 444–471. [CrossRef]
22. Alcock, I.; White, M.; Wheeler, B.; Fleming, L.; Depledge, M. Longitudinal effects on mental health of moving to greener and less green urban areas. *Environ. Sci. Technol.* **2014**, *48*, 1247–1255. [CrossRef] [PubMed]
23. Berman, M.; Jonides, J.; Kaplan, S. The cognitive benefits of interacting with nature. *Psychol. Sci.* **2008**, *19*, 1207–1212. [CrossRef] [PubMed]
24. Roth-Cerina, M.; Rister, V. The Open-Air School Revisited. In Proceedings of the International Conference 'Making Healthy Cities for People-Education, Research, Practice in Planning, Architecture and Engineering' HURBE2021, Sarajevo, Bosnia and Herzegovina, 4–5 October 2021.
25. Sullivan, L. The Tall Office Building Artistically Considered. *Lippincott's Mag.* **1896**, *57*, 403–409.
26. Lau, S.; Gou, Z.; Liu, Y. Healthy campus by open space design: Approaches and guidelines. *Front. Archit. Res.* **2014**, *3*, 452–467. [CrossRef]
27. Newton, C. Vertical schools on the rise in Australian cities. *Archit. Aust.* **2019**, *108*, 86–91.
28. Nychka, J.; Chen, P. Nature as inspiration in materials science and engineering. *JOM* **2012**, *64*, 446–448. [CrossRef]
29. Kaplan, R.; Kaplan, S. *The Experience of Nature: A Psychological Perspective*; Cambridge University Press: Cambridge, UK, 1989.
30. Architectus. Flinders University Health and Medical Research Building. 2024. Available online: https://www.flinders.edu.au/hmrb (accessed on 7 July 2024).
31. Keane, T.; Linden, T.; Hernandez-Martinez, P.; Molnar, A.; Blicblau, A. Digital technologies: Students' expectations and experiences during their transition from high school to university. *Educ. Inf. Technol.* **2023**, *28*, 857–877. [CrossRef] [PubMed]
32. Zhong, W.; Schröder, T.; Bekkering, J. Biophilic design in architecture and its contributions to health, well-being, and sustainability: A critical review. *Front. Archit. Res.* **2022**, *11*, 114–141. [CrossRef]
33. Almusaed, A.; Almssad, A.; Najar, K. An innovative school design based on a biophilic approach using the appreciative inquiry model: Case study Scandinavia. *Adv. Civ. Eng.* **2022**, *2022*, 8545787. [CrossRef]
34. Laboy, M. Physical environments for active schools: Future directions for transdisciplinary research. *Transl. J. Am. Coll. Sports Med.* **2019**, *4*, 155–164. [CrossRef]
35. Baker, L.; Bernstein, H. *The Impact of School Buildings on Student Health and Performance: A Call for Research*; The Centre for Green Schools and McGraw-Hill Research Foundation: Washington, DC, USA, 2012.

architecture

MDPI

Article

The Biophilic School: A Critical Synthesis of Evidence-Based Systematic Literature Reviews

Kenneth Fisher

Melbourne School of Design, University of Melbourne, Parkville, VIC 3052, Australia; fisherk@unimelb.edu.au

Abstract: Biophilic design emphasises human connections to nature to enhance health and well-being. In health and medical environments, the biophilic design hypothesis suggests improved healing and health outcomes when patients have a connection to nature. Emerging evidence indicates that if teachers and students experience a greater sense of health and well-being in the campus environment, they are likely to have better teaching and learning outcomes. To understand this at a deep level, a critical synthesis and analysis of some fifty systematic reviews outlines the significant emerging body of evidence in support of this approach. Systematic reviews are a fundamental basis for new research to be framed in the context of existing research findings within the evidence-based research methodology. Key outcomes of this critical synthesis review are that there is compelling evidence of some 70 health and well-being factors being improved though biophilic design. Also, organising an agreed taxonomy of biophilic design and health and well-being elements at a granular level would assist in simplifying and compiling the evidence of the benefits of such an approach to inform school funding, planning, design and operations authorities.

Keywords: biophilic; school; design; health; well-being; evidence-based; systematic literature review

1. Introduction—Well-Being, Health and the Environment

The connection of health and wellbeing to nature has been increasingly studied in health and hospital planning and design over the past two decades [1,2]. Only recently, however, has such an approach been applied to school planning and design [3].

In undertaking this extensive synthesis and analysis of ss (systematic literature reviews), it became apparent that there is a need to reflect on the lineage of biophilic developments since Fromm [4] framed the biophilia hypothesis concept. In 'The Anatomy of Human Destructiveness', Fromm despaired of humankind's treatment and relative ignorance of the critical role nature plays in our lives and in the current epoch known as the Anthropocene. The biophilic hypothesis was developed further by Wilson [5], and subsequently by Kellert and Wilson [2], and more recently by Kellert [6] and Kellert and Calabrese [7]. Such an approach in school planning and design has had some limited exposure [8], and this study refreshes that approach and takes a much stronger account of a human-centred focus on school design through an evidence-based biophilic planning and design paradigm shift.

The past decade or so has seen a significant focus on health and well-being, as illustrated by the Australian Government's Department of Health and various other State government departments across Australia [9]. This study examined 'Health and the Environment: a Compilation of the Evidence'. The links between public health and the environment are also becoming subjects for significant research output, as seen in the RMIT University Healthy Liveable Cities Laboratory [10]. Recently, the South Australian Government has named its health organisation the Department for Health and Wellbeing.

2. Evidence-Based Health and Education Environments Planning and Design

In terms of evidence-based health planning and design, a milestone was achieved when a team of researchers led by Ulrich conducted a systematic review of all the evidence related to the rate of healing of patients connected with the quality of the hospital physical environment [1]. A team of researchers produced a seminal paper analysing some 223 peer-reviewed papers on the topic demonstrating the connection between positive health outcomes with exposure to nature. This also connects closely with the International WELL Building Institute [11] which focusses on design for well-being.

Designing schools for health and well-being using the concept of 'flourishing', salutogenic [12] and eudaemonic design [13] are illustrated in Figure 1. Flourishing is a cross-disciplinary focus linking a wide range of factors, as illustrated in Figure 2 [14]. All these factors should be considered in school salutogenic and biophilic planning and design processes.

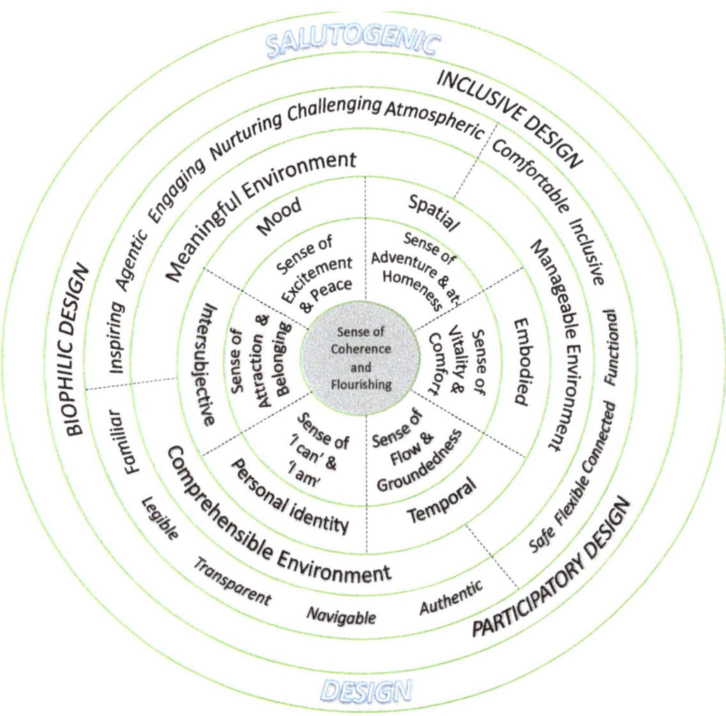

Figure 1. Salutogenic design framework for well-being as flourishing [13].

Biophilic design elements are covered in the literature and case studies but are organised—and to some extent simplified—in different ways by various authors such as Kellert and Calabrese [7] with 24 elements, Browning [15] with 14 elements and Lei [16] with 9 elements. These can be compared with the ever-evolving WELL Building Standard [11] and the World Green Building Council [17]. The interconnections and relationships with all five approaches can be seen in Figure 3.

The key simplified emerging themes included psychological; neurodiverse; physical; performance; emotional; health and well-being; and biophilic features. The search resulted in repeated examples of the impact of the environment on various aspects of health and well-being. For example, cognitive appeared in 16 articles, restorative in 13 articles and so on. These themes and the elements within each theme are summarised in Table 1.

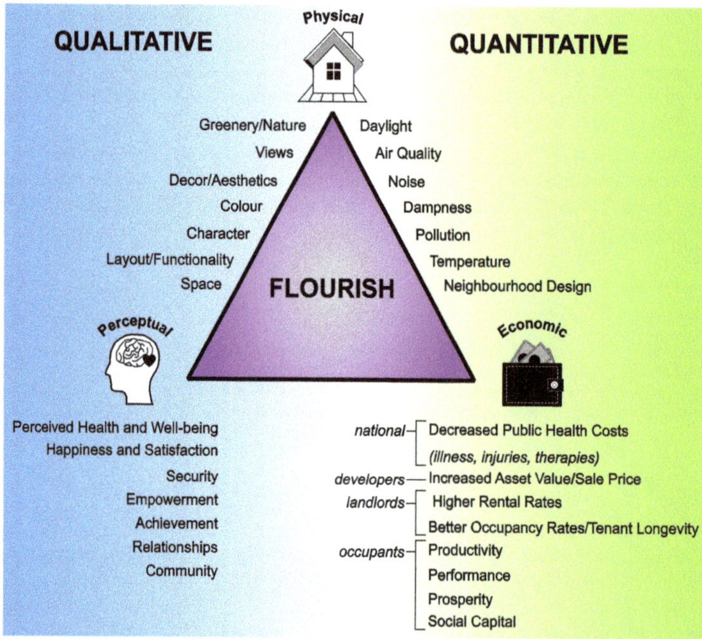

Figure 2. Flourish Model [14].

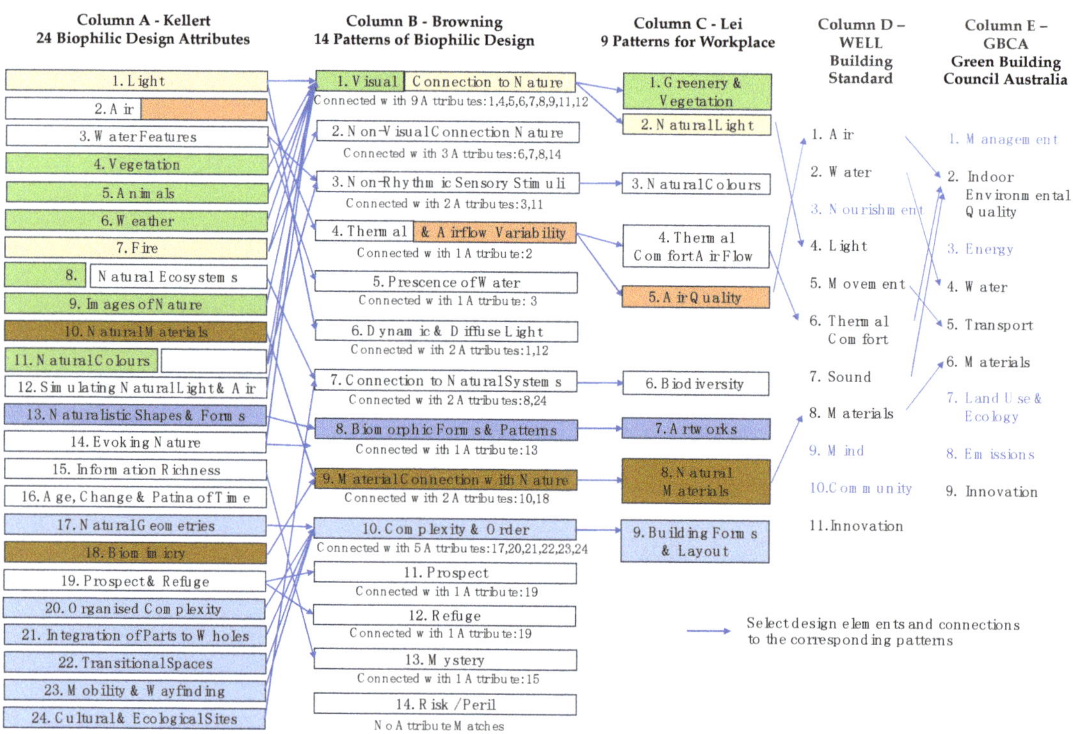

Figure 3. Just some of the biophilic models on offer [7,11,15–17].

Table 1. Categories of Key Abstract Words searched in shortlisted reviewed articles (generated by author).

Impact on Health and Well-Being	Elements in this Category in First Scoping Review
Cognitive, Psychological and Mental	Sensory; Attention Restoration Theory (ART); Stimulation; Psychiatric; Disorders; Cortisol; Sensing; Anxiety; Consciousness; Cognitive; Mental; Behavioural; Psychological; Innate
Performance and Intellectual	Performance; Attentional Performance; Desire; Success
Physical and Physiological	Restorative; Stress; Comfort; Proximity
Emotional and Mood	Identity; Affect; Mood; Connectedness; Perspective; Personal growth
Health, Well-being, Social, Behavioural and Developmental	Therapeutic; Flourish; Fatigue; Trauma; Healing; Health; Well-being; Electroencephalography (EEG); Holistic; Safety; Eudaimonic Well-being; Hedonic Well-being; Morbidity; Healthy; Literacy (health)
Neuroarchitecture and Neurodiverse	Neuroeducation; Neurodiverse; Neuroscience; Neuroarchitecture; Spectrum; Autism; Autism Spectrum Disorder (ASD); Functioning; Disability
Biophilic Features	Biophilic Quality Index (BQI); Nature; Trees; Indoor Air Quality; Greenspace; Empirical; Outdoor; Pollution; Green walls; Greenness; Ecosystem; Landscape; Interiors; Environments; Patterns; Meta-analysis; Bush Kinder; Ethnography; Systematic; Climate; Plants; Sustainability

3. Methods

This analysis and synthesis of multiple literature reviews first conducted a scoping literature review followed by an analysis of the evidence-based articles available from reputable peer-reviewed publishing venues. Systematic reviews are the highest form of scientific evidence with the methodology controlled through the PRISMA process (Preferred Reporting Items for Systematic Reviews and Meta-Analyses). A synthesis of multiple systematic reviews analyses the 'eligibility criteria, identification and selection of studies, data collection and study appraisal, and synthesis and findings' [18]. Tools such as ROBIS (risk of bias) and AMSTAR-2 (methodological quality) are used to filter the data.

Some 1250 articles were sourced from venues as outlined in the PRISMA diagram in Figure 4 using the key words 'school; university; campus; health and wellbeing and biophilic planning and design' over five years. The 1250 articles were screened and then shortlisted as per the PRISMA process. That scoping literature review resulted in the shortlisting of 183 peer-reviewed journal articles and book chapters on biophilia in schools. This was then reduced to 67 articles which, collectively—when the Abstracts were searched for keywords—resulted in approximately 70 keywords.

Systematic literature reviews are a form of 'expert elicitation' [19]. This approach operates in a cross-disciplinary fashion and assists in aligning concepts, language, and evaluation methods without being focussed on any pre-determined paradigm. In short, if many experts can agree on a particular concept—especially in a qualitative narrative way—then the concept may be seen as being valid. Within the final 67 articles were 47 systematic literature reviews plus several related studies of the impact of nature on health and well-being which included workplaces and several clinical settings. There is increasing associated evidence within these sectors which may reflect or relate to findings in the school as a workplace. Having analysed all the final selected SLRs and additional relevant articles, it was decided to settle on the following themes: (1) psychological, cognitive and academic performance; (2) child development, early learning and forest schools; (3) greenspace and outdoors; (4) neuroarchitecture and neurodiversity; (5) higher education sector; (6) general health and well-being; and finally, (7) residential, clinical, health and workplace. These all have implications as an evidence base for supporting biophilic school design for health and well-being.

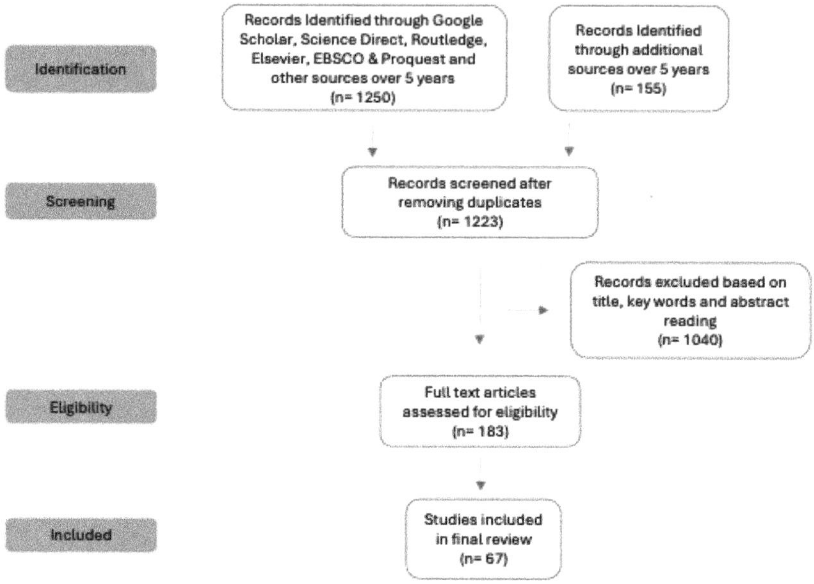

Figure 4. PRISMA flowchart.

4. A Critical Synthesis of the Systematic Literature Reviews

Note that all the references within each of the systematic literature reviews are not individually cited as it would make this paper too long. However, sources of any statements hereunder can be found within those individual systematic reviews that have been cited.

4.1. *Psychological, Cognitive and Academic Performance*

Some of the highlights in this category include improved cognitive development if the school has increased tree canopy [3]. Blood pressure and stress are also reduced in biophilic environments, and learning outcomes improve with greater natural light [20], which is also associated with increased physical activity, subjective well-being, improved alertness, faster cognitive speed, and better concentration. Green environments adjacent to campuses were also found to improve academic performances [20].

Views and sounds of water assist in restorative effects [21], and this was physiological rather than psychological. Indoor plants can also result in improved performance, with various options offering alternative benefits [22]. Views of nature provided varying benefits depending on the type of nature in view. In this context, attention restoration theory (ART) suggested that in some cases artificial images can be more beneficial. Natural materials can also be effective in restoration, although research is limited. The use of timber is perhaps considered to be the most effective.

Stress reduction and mental restoration are likely to improve with connection to nature, whilst students exhibit significantly improved performance on attention test and stress recovery [22]. Connection to nature also sees improvements in other areas including creativity, depression, PTSD, attention deficit recovery, happiness and life satisfaction, enhanced memory retention, dementia, productivity, and self-esteem.

Attention restoration theory [23] and the influence of nature on cognition and stress recovery [24] through a psycho-physiological engagement with nature is used to argue that nature improves academic performance [25]. Access to nature improves concentration over longer periods of time, and the immersive effect improves focus and energy levels. However, only two studies focussed on schools specifically. One of these used data from the Twin Cities Metropolitan Area (TCMA) of Minnesota, which has data on educational performance modified for social and economic disadvantages. It also has data on vegetation coverage and, after discarding schools that were confounding, resulted in a sample size of 222 schools with 3rd grade students being scored using the Minnesota Comprehensive Assessment of the Department of Education for reading and mathematics. Both measures showed improvements after regression analysis.

Nature in the study environment was analysed in a systematic literature review by van den Bogerd [26]. This study developed a three-tiered nature impact arrangement for evaluation: (1) reducing harm (mitigation); (2) restoring capacities (restoration); and (3) building capacities (instoration). Note that 'instoration' implies an active engagement with nature, whereas restoration can be 'passive' engagement. The reviews included 14 studies in secondary schools and 23 studies in universities, with campus green space examined together with the impact of indoor plants. No differences were found between the two educational sectors. Campus green space was found to improve restoration and general well-being and quality of life.

Greenspace was also found to reduce temperature on campus grounds. The studies also differentiated between passive and active exposure to nature. Furthermore, inclusion of water and views improved the experience of indoor nature and related health measures. In some studies, there were issues in bias, confounding factors and baseline differences, and more longitudinal and controlled evaluations were recommended in future studies. The studies reviewed were European-focussed, and it was suggested that they would not necessarily directly translate to other countries.

Another systematic review [27] found that there is evidence to support that exposure to nature improved cognitive performance regarding selective attention, sustained attention and working memory, together with enhanced well-being, cognitive restoration, and stress reduction. Additional factors included improvements in emotional and behavioural issues around inattention and hyperactivity.

Testing included impulse control, processing speed, long-term memory and knowledge acquisition, selective attention, sustained/selective attention and working memory. Most of this work used passive nature attributes. They also noted that in establishing the quality of articles they used the EPHPP (Effective Public Health Practice Project) framework, which assesses six domains—selection bias, study design, confounders, blinding, data collection methods, withdrawals, and dropouts. The study also concluded that nature exposure type—passive or active—is critical in framing any research programme.

4.2. Child Development, Early Learning and Forest Schools

Whilst Adams and Savahl's [28] study focussed more on perceptions of and attitudes toward nature, there are some findings regarding active engagement with nature which can improve cognitive, physical, affective effects, and moral development. Articles included illustrated how nature is important in healthy child development, suggested subjective well-being, provided a buffer to life stress and cognitive functioning, and noted the variable impact that nature can have on cognitive styles. Nature exposure was also thought to mitigate ADHD symptoms.

Arola et al. [29] separated qualitative and quantitative articles. They found that nature connectedness showed a decrease in psychosomatic symptoms and increased self-satisfaction, positive mood, and social interaction. Other factors which showed improvement included reduced stress; increased happiness and joy; improved mindfulness and spirituality; greater competence, self-esteem and emotional well-being; enhanced social well-being and competence; and exhibited an overall heightened sense of physical well-being.

The measurement and categorization of natural environments is key to evaluation [30]. This particular study discriminated between outdoor spaces as 'natural environments' and indoor spaces as 'virtual environments' because of the built form of the latter. Various datasets were used in this study to measure natural environments. Regarding health outcomes, six categories were evaluated in the selected articles: academic achievement;

prevalence of doctor-diagnosed disorders; emotional and behavioural functioning; well-being; social functioning; and cognitive skills. Emotional and behavioural function included attention, interpersonal relationships and ADHD. Cognitive skills assessment used a test for memory performance and cognitive performance. All test tools used were listed across the six categories. Academic achievement was seen to improve if exposed to natural environments and doctor diagnoses were lower. Behavioural and emotional aspects were improved and there was also improvement in some aspects of memory.

Another related study long listed massive numbers of articles—some 37,159 references reduced to 65—with the final articles showing various improvements in a range of child development factors) [31]. The review is quite complex but does suggest strong improvements in nature play's impact on social skills, independent thought, lower behavioural problems, emotional development and well-being, cognitive development and cognitive play, creativity, imagination and dramatic play and self-regulation.

Regarding the psychological impact of nature on 6–12-year-olds, it has been suggested that there are more cross-sectional studies than longitudinal ones, with an equal number of observational (quantity and accessibility of green spaces) and experimental studies, but little work on blue/water impact [32]. In evaluating the quantity vs. quality of nature/green spaces, it was found that observational and quality studies can result in positive affective responses including joy, excitement and restoration. But some nature features can manifest stress, danger and threat feelings. Experimental studies showed that views of both nature and simulated nature had a positive impact on well-being, but that actual natural engagement had a higher impact. Active engagement in urban forest settings can improve psychological well-being, self-esteem and peer relationships. Overall, the systematic review revealed a positive impact on psychological well-being with increased engagement with nature.

An experimental study using the notion of personal comfort systems (PCS) is included here in large part because of this author's personal experience in schools regarding thermal comfort [33] The PCS model was used with 18 respondents in a controlled environmental room—airspeed, temperature, relative humidity and lighting were controlled—with the addition of being able to cool the headspace and extremities of individuals at a desk setting with various user controls. Fourteen temperature sensors were applied to the body. Various tasks were undertaken as specified within several professionally accepted frameworks.

Tasks included anti-cue, complex tasks, planning tasks, verbal ability, working memory and mental spatial manipulations. Measures included NASA-defined subjective perception of cognitive load task-induced physiological responses, arousal and alertness and interaction between tasks and performance. Outcomes included the finding that cooling the head (if ambient is over 25 °C) improved the anti-cue task and complex cognitive performance. Lower temperatures also increased task performance. Conversely, however, warming from a colder temperature did not affect complex cognitive performance.

Meng's systematic literature review of the indoor visual environment of children selected five environmental factors for analysis including: lighting, access to nature, window features, art and environmental aesthetics, and ergonomics in spatial settings [34]. It found that appropriate lighting and access to nature had the most evidence supporting benefits to children's health, with dynamic lighting responding to children's activities being the most beneficial.

A number of studies had some evidence of possible bias, which raised questions about the quality of the research methodologies. The lighting studies feature heavily in the United States, whilst the nature-restorative studies had a greater representation in Europe.

Russo [35] found that access to nature can elicit positive multiple childhood health outcomes, including decreased emotional and behavioural difficulties, higher academic achievement and improved overall mental well-being and cognitive development. Another survey of 145 schools in Scotland by Robertson et al. [36] saw 82% of respondents say that it is vital/essential that students should spend time outdoors regularly; 85% of schools say students use the outdoors every day; 47% said that students should visit woodland regularly. Whilst only 30% of schools have woodland on their campus, 78% can access nearby woodland, with between 66% and 80% accessible by walking.

That said, only about 43% visit the nearby woodland site regularly. Schools ranked in importance the following benefits generally regarding access to nature as: Learning Opportunity (23%); Natural Resources (22%); Emotional Well-being (17%); Physical Activity (16%); Better Quality of Play (15%) and Better Behaviour (8%). There was strong interest in developing a forest kindergarten, with up to 68% of organisations wanting to do so 'as soon as possible'.

Forest schools provide excellent opportunities for deep research in the connection between nature and health and well-being [37]. Interaction patterns (IP) in 'wild natural' parks were used to develop a human–nature interaction model which was then used on two outdoor classrooms in a forest school located in an arboretum. After 35 weeks of filming interactions in six zones in the outdoor classrooms, these interactions were coded and analysed. From 1,851 interaction patterns, 25 were determined to be keystone patterns, with 8 of these being selected as indicative of bonding with 'wild' elements of nature. These included cohabitating with a wild animal, imitating animals, calling birds, leaning against a tree, lying on earth, being in solitude in nature actively, and being in solitude in nature.

Key findings included that in outdoor classrooms, active learning serves to strengthen child bones, muscles, hearts and lungs. It also improves coordination, balance, and posture and helps in sustaining a healthy weight. Moreover, it reduces illnesses including cancer and diabetes. Engaging with 'wild' nature suggests that students are in a relational connection with nature, rather than dominating nature.

4.3. Greenspace and Outdoors

Observational studies of the relationship between greenspace on mental health are instrumental in much of the work in this field, with 'blue space' making up 25% of studies [38]. Generally the quantity of greenspace was used in longitudinal studies exploring mental health and it was recommended that further research should also focus on the quality of greenspace. Studies were carried out in Europe, North America, Asia, Australia, New Zealand and South America, with the UK contributing 24% of all studies and Japan contributing 15% of experimental studies. Scale was a significant issue arising out of these studies, both in terms of the scale of nature and the scale of the study respondent numbers and their variable time spent in nature and greenspace.

Whilst both observational and experimental studies were studies, it was found that the experimental studies are viewed as more 'gold standard' than observational studies. The review could not establish a direct causal relationship between greenspace and mental health because of the variations between the scale and quality of greenspace, the variations in time spent in the space and the varying longitudinal nature of the studies. These factors need to be considered in future studies.

Conversely, a meta-analysis of prescribing a rehabilitation approach using nature-based interventions to improve the depression of community-based adults [39] found that an optimal dose of 12 weeks of gardening, green exercise and nature-based therapy can improve mental health problems in adults (perhaps more suited to senior secondary school in this context), including for those with pre-existing conditions. In particular, random controlled trials showed a reduction in symptoms of depressive mood. Forest bathing proved the most effective in this regard and reduced depressive mood symptoms. There was also a large decrease in anxiety symptoms, together with a large increase in positive affect. Gardening was seen to have a significant effect on physical activity over a year-long study with respondents who had serious mental illness.

A narrative synthesis study by Fernandez of 39 research articles focussing on 5- to 12-year-olds considered three factors: indoor environmental quality (IEQ); increasing student 'green time' by school greening; and more active travel to school [40]. The review suggested that the studies were not sufficiently robust to be considered useful. That said, it did find that greening schools does influence cognition and physical activity, although behaviour was not considered to be affected at least in this study.

Another study used student academic performance as an indicator of cognitive abilities connected with or in close proximity to green spaces [41]. Three green indicators/metrics were used: (a) GIS/NDVI (Geographic Information System mapping technology and Normalised Difference Vegetation Index, a satellite indicator of green canopy density); (b) distance to green spaces (buffers of 500, 750 and 1000 metres around schools); and (c) the quantity of green spaces. Academic performance was sourced from Government Education databases. Whilst the study did find a relationship between the distance of greenspace and academic performance, the quantity of greenspace was a key factor which needs further consideration.

4.4. Neuro-Architecture and Neurodiversity

Neuroscience and architecture are becoming a significant feature of research focussing on the connections between the two. Neurodiversity is also of increasing importance for students with learning difficulties requiring much greater attention as to how the physical learning environment might impact on their capabilities.

Four aspects of well-being are the focus of one such systematic review—the physical (body); the intellectual (brain); the emotions; and the social (behaviours) [42]. The study used a descriptive thematic, narrative and critical analysis approach which produced a heuristic neuro-architectural model relating design characteristics with 'experimentally proven' physiological, psychological, cognitive and behavioural effects. The model relates these human factors to architectural designs ('unconscious' perceptions); physiology (human body functions); psychology (emotions and behaviour); and neuroscience (neurology, cognition, perception).

Various physiological measurement tools were used in differently designed environments, with both physical and virtual experiences. Few studies considered smells and dark colours, but natural materials such as wood led to lower heart rates and less sweating in both environments. The authors noted that physiological changes (both subconsciously and consciously) related to different architectural features altered blood flows to different parts of the brain. The resulting emotions were thought to modulate the brain activity in areas related to attention, memory, motivations, emotions (mood) and decision-making which all led to better psychology, well-being and restoration.

Another systematic review studied interior architectural designs and their impact on the autonomic nervous system (ANS) and/or the central nervous system (CNS) [43]. The study separated affect and emotion, with the former related to valence and arousal and the latter being defined in neurobiological terms as a complex reaction stimulus of cortical brain networks. Blood oxygenation was measured as an indication of neuronal stimulation. In addition, heart rate, blood pressure, breathing rate and galvanic skin responses were measured. The physical interiors were organised in a confined floor and ceiling space having four sides. Greater than four sides and curvilinear surfaces were considered to be semi-enclosed if there were solid partial barriers. Findings included that lower ceilings resulted in a lower judgement of beauty; linear geometries resulted in lower self-ratings for pleasure and arousal, with decreased anterior cingulate cortex activity. Furnished interiors had higher self-rated emotions related to presence and arousal according to the style of furniture, with increased heart rates. Depth in virtual environments increased the sense of presence, and wood (as materiality) reduced the impact on the ANS.

Cognitive judgements of architecture can also be a response to prior external experiences rather than internal emotions [44]. In this context, 'extrospective' and 'introspective' evaluations can be related to 'dissociable neural circuitry'. Five cognitive architectural judgements were considered, namely: complexity; organisation; modernity; naturalness; and beauty. These characteristics can affect the degree of physiological stimulation experienced. They can be uplifting or depressing, and these responses can predict long-term health measures including stress and depression. Vitality and valence (the latter meaning whether the space is desirable or undesirable) are related to preference, liking and pleasantness. These affective responses relate to neural networks which regulate pleasure and emotion.

For example, 'hominess' ratings can be a result of culture, background and memories which may influence cognitive and sensory processes. There are two characteristics: emotional responses (of which eight are considered including: personalness, hominess, relaxation, comfort, stimulation, uplift, vitality and valence) and behavioural–emotional responses. The latter consist of measures of behaviour, movement and motivation, all of which may be linked to sensorimotor processes in the brain. Three behavioural–emotional responses are considered here including: interest, approachability, and exploration. These can all result in sensorimotor, cognitive and emotional responses.

All of these aesthetic/response measures in this case study are related to, and examined through, three key physical variables, namely: ceiling height; enclosure and curvature. The authors then construct a psychometric network analysis of elements. By a complex process—a little too complex and dense to describe in detail here—the authors then examine responses to three criteria with MRI (magnetic resonance imaging) brain scans namely: hominess; coherence; and fascination. They conclude that open spaces scored more highly. A high ceiling, openness and curvature are also considered to be more beautiful and are more likely to stimulate movement and exploration similar to the previous study [43].

Respondents also preferred greater affordances of visual prospect, and the preferences for high ceilings and curved interiors resulted in more active 'neural structures associated with visuospatial exploration and attention'. Certain visual features could also modulate psychological responses related to fascination. However, they 'suspect' that in some cases, these responses may be more related to extrospective rather than introspective forces. IT was suggested that the three spatial elements could possibly be linked to the 16 aesthetic/response measures in further MRI studies to see if these elements could be added to the abovementioned factors within the notion of flourishing.

Karakas considered the differences between 'theoretic' frameworks and experimental studies [45]. The experimental articles resulted in five themes as illustrated in Table 2. These themes all had associated neurological responses. Many—if not all—of these responses resonate with other systematic literature review findings in this synthesis overview.

Table 2. Relationships between spatial typologies and human responses (by author).

Spatial Theme/Activity	Human Response
Restorative and stress reduction affordance	Well-being, restorativeness and stress reduction
Aesthetic judgment and appreciation	Pleasure, familiarity, novelty, comfort and pleasantness
Pedestrian experience, navigation and wayfinding	Mediation, attention, anxiety, displeasure, positive and negative emotions, intense and mild, pleasant and unpleasant, directional behaviour, familiarity and fear
Visual engagement, visual attention and imageability	Visual attention, avoidant behaviour, conscious and unconscious attention, efficient, pretty, safe, pleasant, interesting
Phenomenological experiences, experiential intensity, user experience, multisensory experience and natural experience of architectural spaces	Relaxation, excitement, engagement, reduced stress, focus, interest, attention, appreciation, peace, beauty, connectedness, reduced anxiety, pleasure, motivation, pleasure and frustration

Colour and the perceived 'temperature of colour' are not covered to any great degree in the literature on biophilia [46]. This study has been included to provide some input to afford a more complete biophilic hypothesis. It investigates the effects of personal control of lighting on visual and thermal responses, together with affect and cognitive performance. Sixteen participants were tested with user control of colour-correlated temperature (CCT) of ceiling-mounted lighting, task lighting and computer screen lighting. Prior to this, respondents had a 24 hour-controlled environment, sleeping and eating pattern. Also, prior to the test, various sensors were applied to the individuals.

In the first 1.5 h session, the tasks were applied without any control of CCT. In the second session, respondents carried out cognitive tasks for one and a half hours but could use the CCT controls to vary the settings to their preferences during the first 10 min. They then carried out similar activities to the first trial, but with their own lighting settings. Findings were that thermal appraisal was not significantly affected by having CCT control. However, personal control of CCT improved visual acuity and also mitigated eyestrain, although it did not improve cognitive performance.

Whilst Parhizkar [47] focussed on office workers, findings in this sector can be relevant to schools in many ways. That study used a Flourishing Index (FI) to assess the well-being of office workers in relation to indoor environmental quality (IEQ). Note that 574 office workers located in 7 separate buildings were surveyed using a 5-point scale. The FI scores correlated with IEQ satisfaction, including noise, temperature and light. With higher CO2 and lower IEQ satisfaction, lower FI scores were related to reports of respiratory symptoms, shortness of breath, coughs and sore throats. The connection between CO2 and FI could be mediated by improved IEQ factors, but PM2.5 (particulate matter in microns, a key issue in Bangkok schools) directly affected FI ratings. The flourishing index rated happiness and life satisfaction, health, meaning in life, character strengths and social relationships.

Environmental psychology has also provided a rich vein of research on neuroarchitecture and biophilic design [48]. Joye argues that psycho-evolutionary development can be seen as a framework and can be used to track how humans have evolved in terms of their relationship with nature and its impact on cognition. This suggests that different environments can have differing impacts on our affective experience of those environments. Kaplan [23] and Ulrich [49] differ in that the former argues that we respond to information we receive in each setting, whereas the latter argues that we have an instinctive flight or fight response which dates back millennia (see below). The notion of affective valence supports the latter concept.

Regarding the neural origin of these affective states, some researchers attribute an important role to subcortical areas, especially the amygdala. Because these structures are also involved in modulating stress-related hormones, it provides an explanation of why certain types of settings have a different influence on autonomic stress responses [48].

In relation to the attention restoration theory (ART) discussed earlier, Joye compares Kaplan and Ulrich's interpretations. The former argues that ART is related to attentional capacities, whereas the latter believes it is related to stress, which can occur while attentional capacities may not be impeded. So, whilst the flight or fight reaction might be useful, it can still result in psychological and physiological stress. There are variations in these reactions, depending on the type of landscape being experienced, which Ulrich calls Structural Landscape Features. Biophilic design attempts to fold these structural landscape features into the building designs, so that the restorative effects can be experienced.

It is through this model that some of Browning's 14 elements are derived, such as prospect, refuge, mystery, curiosity, coherence and complexity [15]. Later developments on this idea have led to the transfer of fractal geometry from nature into architectural design features. Joye discusses the links between various natural fractal geometries, restoration and preferences, which Ulrich calls the affective value of fractals [24]. Joye describes various examples of the use of fractal architecture deploying repeated scales using gothic architecture and Hindu temples as examples.

Nature-based architecture implies that the building enters a dialogue with a specific set of human inborn affiliations. However, adherents of biophilic architecture should become aware that their work also has to relate to or become embedded in a social, historical, ecological and individual context [48].

Another neuro-architectural study notes that in the developed world, people spend as much as 90% of their time indoors and, in some cases—particularly in the UK, for example—as much as 95%, meaning that connection to nature is minimal [50]. Valentine asserts that physiological stress can be linked to various diseases including cardiovascular, neurodegenerative, autoimmune fatty liver disease, chronic kidney disease and cancer.

This systematic review found that physical design attributes which can impact affect and preference include curvature, enclosure and proportion, as discussed earlier. Valentine suggests that clinical biomarkers can be used to measure the impact of these features on stress.

These markers were measured using electroencephalography (EEG); functional magnetic resonance imaging (fMRI); heart rate variability (HRV); galvanic skin responses (GSR); pupil dilation; and salivary cortisol. Three forms of architecture were used—virtual reality (VR); 2D images; and physical reality, i.e., actual buildings. The three abovementioned forms were all tested, and the responses were measured. Some of the examples used classrooms, but the author suggests that the variety of spaces used made it a complex task to compare results and suggested that a more standardised research methodology model should be developed. The ultimate findings are that the biomarkers are a valid way of measuring responses, and that the three spatial typologies—with all their complex variations—can influence these biomarkers.

4.5. Higher Education Sector

Several reviews focussed on university campuses. Abdelaal found a connection between nature and cognition, cognitive performance, reduced mental fatigue, improved productivity and attention, academic performance, attentiveness, improved concentration and memory [51]. The study categorized three types of restoration: (a) physical and psychological stress reduction (five references); (b) psychological self-esteem, mood, anxiety, attitude, and happiness (seven references); (c) social interaction and spiritual well-being (six references). The study looked at two types of biophilic interaction—degree of interaction and interaction features and patterns. The health and well-being impacts were measured in four categories—physical/physiological, psychological, cognitive and social/spiritual. In terms of retrofitting biophilic features into campuses to improve health and well-being, the author notes several barriers including the existing spatial configuration of campuses and buildings of the past which makes it difficult to retrofit biophilic designs.

University libraries are learning spaces which have had little research carried out in the context of the health and well-being of students and staff. Now Khair's systematic review has found benefits of biophilic design on the performance of academic, intellectual and cognitive tasks, student productivity, creativity and physical activity as well as nourishing intellectual curiosity and innovation [52]. Views of nature also benefit mental breaks (restorative), improved attention function and learning. Key biophilic factors were found to be spatial proportions, natural ventilation, indoor plants and daylight. There was also a calming effect and improved stimulation, wellness and performance.

Students also appeared to have improved emotional, psychological and physiological benefits and were happier regardless of demographics. Relaxation is improved, as is self-esteem, and there were overall improved feelings due to improved blood pressure and heart rate. Visual, auditory and olfactory stimulation all appear to be improved, with floral fragrances thought to enhance better moods. Students also appeared to prefer to study in 'refuge' spaces with 'prospect' views and to feel privacy, security and excitement. Natural light is preferred, but artificial light designed to suit the task and spatial typology were also critical to comfort. Study spaces next to windows were preferred. Vegetation was seen to be an effective influence on microclimate and the moderation of noise levels.

Peters' review of university campuses supported many of the abovementioned findings including that lighting systems serve to make learning spaces attractive and pleasant and enhance the sense of spaciousness, assist in delineating settings, stimulate learning and show better learning behaviour, whilst students in areas of natural light perform better than those in artificial light [53]. In addition, this study noted that a sense of place and place attachment increased, including an enhanced sense of happiness that was evident with biophilic design. Nature connectedness was also reported by students as fostering innovative and holistic thinkers. The preference for 'refuge' study space together with 'prospect' was also supported. IEQ improved student productivity, with adjacency to windows the most effective. Connections to nature also reduced depression and ADHD symptoms. The systematic review also provides guidelines for biophilic design interventions to support these abovementioned enhancements to student experiences and their health and well-being.

Saarani's study of university students focussed on emotion, health and well-being and cognition, with each category of these elements being rated on the numbers of mentions in the review articles [54]. The biophilic impact on emotion ranked highest on attitude, with psychological restoration and positive emotional responses rated next. Self-esteem, feelings of comfort and reduced anger are also indicated as improving. Reduced mental fatigue, illness and anxiety featured in the health and well-being category. Cognitive performance was found to have the most impact in relation to biophilic effects, with improved productivity, commitment and ability to perform, and a positive impact on concentration, memory, mental engagement and attention being the most improved through biophilic design. Additional affective factors which are enhanced include cognitive performance, satisfaction and confidence levels. At a more reduced rate of improvement lie academic performance, inspiration and motivation, social interaction and creativity.

Campus greening is also an area receiving more attention in biophilic research. van den Bogerd's review examined the effects of campus green space on students' well-being and on students' academic outcomes [55]. A higher quality of life was suggested in four studies where there was greater access to—and higher availability of—campus greenspace. In addition, there was seen to be better general health (these two were seen as medium risks). Regarding green space and student academic outcomes, one moderate risk study found that tree canopy within a one-mile buffer was associated with higher test scores. One high-risk study noted that campus green space improved secondary students' knowledge retention after an outdoor classroom activity. Yilmaz and Ayten [56] provide an excellent analysis and case study as to how Browning's [15] fourteen biophilic patterns can be applied to a university campus.

4.6. General Health and Well-Being

There are now many studies on the relationship of nature to health and well-being outside of the educational sector, and the systematic reviews of these examples are useful to compare and add to the school-based studies. One such study around healing landscapes saw architectural students taking photos of the views out of their own homes to discuss in class [57]. The resulting systematic review explored the concept of the 'biophilic hierarchy of needs'—or BHON —which is adapted from Maslow's hierarchy of needs [58,59]. The human impact measures used groupings around physiological, psychological, cognitive and well-being outcomes.

Zhang's integrated narrative review explored the effect of nature on physical and psychological needs [60]. It notes that people find different benefits from exposure to nature and can also gain benefits from very short exposure in some cases. The authors developed a 'new heuristic point of view' on the necessity of these connections. In the case of physical health, they focussed on cardiovascular disease; immunity; synthetic nervous system function; and sleep quality. For psychological well-being, they reviewed mental health; attention; mood; stress; and cognitive development. These were qualified and measured against nature exposure and nature connectedness. They conclude that the varied nature of exposure and connection to nature and the varied benefits require a heuristic or bespoke approach to optimise those benefits.

4.7. Residential, Clinical, Health and Workplace

As noted earlier, other sectors may have relevance to biophilic school design in adding to an increasing evidence base for this practice. It was suggested that there was little evidence concerning cardiovascular disease (CD) and related death rates and connections to nature, and so it was decided to conduct a systematic review focussing on this aspect [61]. Gascon's final 12 studies were heterogenous in design, with similar population sizes, green space evaluation and covariate data. It found conclusive evidence that CD mortality was reduced in areas of higher-density residential green spaces. Gascon suggested that further studies might account for socioeconomic status and that green spaces need to have a consistent definition regarding density and typology.

The workplace is considered to have little research on the impact of nature on workers (contrary to another abovementioned study), and it is suggested that what exists is limited and scattered [62]. Goncalves' systematic review considered outdoor and indoor nature exposure, and this exposure was related to well-being, motivation, job satisfaction and work performance. Of the 16 shortlisted studies, 9 were based on simulated nature exposure in the indoor spaces, and these simulated exposures also had positive effects on the employee factors measured. Likewise, the outdoor spaces also had positive effects. It was suggested that 12% green coverage indoors was an optimal measure to positively impact both psychological and physiological outcomes, including productivity. In quoting Lei [16] Gascon noted that over 12% up to 20% coverage could negatively affect positive change on physiological brain activities.

In a similar vein, Sadick identified the lack of connection between sustainable design and the health and well-being of office workers [63]. To enhance the missing social sustainability of the human–nature interface, this study also sought to understand how nature exposure in the indoor and outdoor environments could enhance workers' well-being in five categories—restoration; motivation; stress reduction; health; and stress coping strategies—which were considered integral to three areas of study: physical, psychological and cognitive. Nature exposure was observed to be beneficial with all five categories, although the three most impacted included stress reduction, restoration and coping.

The experience of nature was seen as breaking the stress cycle of work activities. The study found that indoor exposure is less effective for restoration and stress reduction due to the proximity of the sources of stressors, although it is essential to have indoor interventions to improve motivation and health to sustain cognitive performance. It emphasised the requirement to have a diverse range of nature experiences as people respond to differing exposures in different ways. That is, a 'once size fits all' is not the preferred approach. The study arrived at a network mapping diagram to illustrate the respective pathways of nature exposure and the resulting positive outcomes on the workers.

Tekin's research in a clinical hospital environment noted the lack of an agreed standardised guidance for the benefits of biophilic design on clinical users [64]. In asking the question 'which biophilic criteria are most critical in a clinical therapeutic environment and how do they inform design?', the study found that three distinctive user groups had specific preferences for the nature exposure experience: outpatients (fresh air, light–daylight, thermal comfort, welcoming and relaxing); inpatients (feeling relaxed and comfortable, prospect refuge, security and protection, light–daylight, view); and staff (privacy refuge, quietness).

Many of the studies reviewed were from departments of medicine, health, design and environmental analysis, architecture, forestry and natural resources, clinical, neuro- and development psychology, as can be seen in Tekin's doctorate on the topic [65]. That study found that 'it was confirmed that emotional, mental and spiritual health issues are typically disregarded while the main foci are physical treatment and cost'.

5. Key Findings and Discussion

Notwithstanding the rapidly emerging body of literature on how nature can impact health and well-being, LeFosse noted that despite biophilia having had centuries of human–nature interaction research and design (to whit Indigenous, Arabic, Moorish, Egyptian, Roman, Etruscan, et al.), biophilia still remains an emerging field of research [66]. This study covered 60 years of literature using a 'three-metric' systematic literature review approach consisting of quality, quantity and application. As one of the few systematic literature reviews to encounter biophilic urbanism, it argues for a greater emphasis on a broader and longer-term scale to promote a paradigm shift in city planning. It calls for new concepts of bioregional systems where 'nature plays a key role in ensuring ecological services and citizens' well-being'. It concluded that biophilia (theory), biophilic design (practice) and biophilic urbanism comprise a triad consisting of disciplinary, metric and spatial dimensions which can form a notion of biophilic upscaling, which is suggested as the best way forward.

Regarding 'biophilic architecture', this also implies that fractal architecture should also respond to critical regionalism, where the terroir, cultural history, vernacular, micro-climate, topography, geology, soil and native vegetation should all be considered as part of a fractal ecosystem [8]. In short, designers should be cognisant of the full environmental ecosystem in planning their buildings, with a view to engaging with—indeed enhancing—biodiversity through the 'nature positive' concept [67].

In this context, Wijesooriya argues that environmentally sustainable design (ESD) does not focus sufficiently on human interactions with the built form [68]. ESD focuses more on the building rather than the approach of 'human centred design'. The WELL certified and biophilic design approaches seek to redress this disparity by covering both ESD and the biophilic effect. Wijesooriya suggests that there are excellent opportunities to capitalise on designing for improved outcomes of cognitive, behavioural, health and well-being and emotional impacts.

Independent reviewers of this manuscript suggested that the findings be presented in a tabular format, as illustrated in Vella-Brodrick's study [27]. This is a challenging task because of the complexity of both the biophilic element typologies—ranging from 9 to 24 in Figure 3—and the range of human responses (numbering seventy) to these biophilic elements, as illustrated in Table 3.

After analysing, synthesising and critiquing the SLRs, it was decided to condense the many biophilic elements named in those SLRs to the following 12 elements, namely: (1) connection to nature; (2) biophilic features; (3) natural light and control; (4) views of nature and virtual; (5) natural materials, especially timber; (6) indoor environment quality; (7) campus green space; (8) green adjacent to campus; (9) increased tree canopy/forest; (10) architectural setting; (11) water views and sound; and (12) indoor nature.

Of the human responses to these environments there were 44 mentions of the various psychological, physical, emotional and mental categories across the SLRs analysed, and these are correlated against the particular biophilic element which has been shown to have a connected positive effect. Given the space available in this article, it is not possible to comment on every point in the matrix. However, some highlights are worth mentioning:

- From the point of view of the biophilic elements, connection to nature is the most used term, followed closely by biophilic features. Other terms are sparsely used, although natural light, control of lighting, views of nature/virtual elements, natural materials (especially timber/wood) and indoor environmental quality (IEQ) are also used in a few reviews.
- In terms of 'human improvement' when associated with a connection to nature, the most mentioned impacts are on academic performance and affect. In this category, we also see numbers of mentions of: flourishing, joy, happiness and pleasure; health and wellness; mental health and fatigue; restoration and relaxation; social interaction, stress reduction and subjective well-being.
- Highlights in the generic biophilic features impact categories include an impact on academic performance; creativity, curiosity, innovation and inspiration; flourishing, joy, happiness and pleasure; productivity(student); and self-esteem, regulation and confidence.
- It is also worth noting that academic performance is mentioned multiple times in 10 of the 12 biophilic element categories.

Table 3. Matrix of Findings.

Human Improvement	Biophilic Elements Referred to in the SLRs										
	Connection to Nature	Biophilic Features	Natural Light and Control	Views of Nature & Virtual	Natural Materials Esp Timber	Indoor Environment Quality	Campus Green Space	Green Adjacent to Campus	Increased Tree Canopy/Forest	Architectural Setting	Water Views & Sound	Indoor Nature
Academic performance	5	4	3	3		3	1	5	1			1
Affect (positive)	3		1							2		
Alertness			1									
Anxiety	2	1			1							
Attentiveness	3	2		1	1							
Behaviour, anger	5	1										
Blood pressure, heart rate, sweat		2			1							
Cognitive & intellectual	7	2	1			1			1			
Cognitive speed		1	1			1						
Cognitive play	1											
Competence	1											
Coordination	1											
Concentration	2	1	1									
Creativity, curiosity, innovation & inspiration	2	6										
Depression	2	1										
Doctor reduced diagnoses	4	1										
Energy levels	1											
Emotional mood & development	6	3			1							
Stimulation & excitement	1	1										
Focus/alertness	1											

Table 3. *Cont.*

Human Improvement	Biophilic Elements Referred to in the SLRs											
	Connection to Nature	Biophilic Features	Natural Light and Control	Views of Nature & Virtual	Natural Materials Esp Timber	Indoor Environment Quality	Campus Green Space	Green Adjacent to Campus	Increased Tree Canopy/Forest	Architectural Setting	Water Views & Sound	Indoor Nature
Flourishing, joy, happiness and pleasure	3	4	1	1		1				1		1
Health & wellness	3	2	2	1		1	2					
Hyperactivity reduction /ADHD	2	1										
Imagination	1											
Impulse control		1										
Independent thought	1											
Life satisfaction	1	1										
Memory, retention	1	2			1		1					
Mental health, fatigue	7	2										
Mindfulness & spiritual wellbeing	2											
Moral development	1											
Motivation	1	1			1							
Movement & exploration	1	1								1		
Psychosomatic symptoms												
Productivity (student)	1	3										
PTSD		1										
Quality of life							1					
Restoration & relaxation	5	2		1	3		1			1	1	

Table 3. *Cont.*

Human Improvement	Biophilic Elements Referred to in the SLRs											
	Connection to Nature	Biophilic Features	Natural Light and Control	Views of Nature & Virtual	Natural Materials Esp Timber	Indoor Environment Quality	Campus Green Space	Green Adjacent to Campus	Increased Tree Canopy/Forest	Architectural Setting	Water Views & Sound	Indoor Nature
Self-esteem, regulation & confidence	3	4										
Social interaction	4	1										
Stress reduction	7	1								1		
Temperature campus							1					
Wellbeing (Subjective)	5		1		1		2					

The numbers of mentions in the SLRs noted above suggest a strong positive connection between various biophilic elements and resulting positive human responses. It has been suggested that some of the measures of these human responses are objective, and some are subjective. Each of the SLRs has notes on which of their identified human responses are measurable. Apropos subjective assessment and 'measurement', it is also possible to evaluate this through the concept of 'expert elicitation' using the IDEA protocol—investigate, discuss, estimate and aggregate [19]. It is clear that there is a significantly increasing amount of evidence-based peer-reviewed research that supports a strong link between many physical, psychological, spiritual and medical benefits in the workplace.

Thus, it can be said that this emerging research in school design illustrates that if schoolteachers and students are exposed to effective biophilic design and biophilic urbanism planning and design elements, they will have an enhanced sense of health and well-being, and many indicators in physical, psychological, spiritual and medical realms can be improved. Related studies discussed above regarding the openness, ceiling height, volume and curvature can also have a positive impact on health and well-being. This is completely at odds when the design of a standard classroom is considered, some 150 years on from when the concept was first developed.

The research is complex and cross-disciplinary, using mixed methods across many sectors of the built and natural environment. Furthermore the 70 or so health and wellbeing factors mentioned across the SLRs makes comparative analysis difficult. The significant amounts of relevant research—perhaps some 600 or so peer-reviewed papers being analysed in the SLRs alone—is very encouraging. This is becoming to look like sufficient quality evidence of the positive connection between nature and health and well-being in schools to be placed in front of policymakers and treasuries to provide more focussed funding to make schools biophilic models of healthier places of learning.

Such a biophilic approach would show leadership to the whole community that schools could lead the way in biophilic design for health and well-being and can demonstrate what is possible in all workplace sectors, including homes and the public realm.

6. Conclusions and Further Research Directions

If we critically scrutinise the school-oriented synthesis of systematic literature reviews analysed in this manuscript, the individual papers referenced and cited within each SLR review analysis need to be summarised and assessed for their methodological thoroughness and validity. This would mean that the total of 600 or so articles referred to within the systematic reviews discussed above would need summarising and ranking for methodological validity. Ulrich had a team of eight people who contributed to his biophilic healthcare planning and design project.

Such an approach may be the next step in the process of developing a robust case for increasing the biophilic nature of school design to enhance health and well-being. Meanwhile, this critical synthesis and analysis of evidence-based systematic reviews considered in this paper can be seen as an interim step in seeking scholarly and educational sector support for such an approach.

Funding: This research received no external funding.

Informed Consent Statement: Not applicable.

Data Availability Statement: Data are contained within the article.

Conflicts of Interest: The author declares no conflicts of interest.

References

1. Ulrich, R.; Zimring, C.; Xuemei, Z.; DuBose, J.; Seo, H.; Choi, Y.; Quan, X.; Joseph, A. A Review of the Research Literature on Evidence-Based Healthcare Design. *HERD Health Environ. Res. Des. J.* **2008**, *1*, 61–125. [CrossRef]
2. Kellert, S.; Wilson, E. (Eds.) *The Biophilia Hypothesis*; Island Press: Washington, DC, USA, 1993.

3. Determan, J.; Akers, M.; Albright, T.; Browning, B.; Martin-Dunlop, C.; Archibald, P.; Caruolo, V. The Impact of Biophilic Learning Spaces on Student Success. 2019. Available online: https://www.brikbase.org/sites/default/files/The%20Impact%20of%20Biophilic%20Learning%20Spaces%20on%20Student%20Success.pdf (accessed on 27 June 2024).
4. Fromm, E. The Anatomy of Human Destructiveness. Holt, Rinehart and Winston. *Horizons* **1975**, *2*, 153.
5. Wilson, E. *Biophilia*; Harvard University Press: Cambridge, MA, USA, 1984.
6. Kellert, S. Building for Life: Designing and Understanding the Human-Nature Connection. *Renew. Resour. J.* **2005**, *24*, 8.
7. Kellert, S.; Calabrese, E. The Practice of Biophilic Design. 2015. Available online: www.biophilic-design.com (accessed on 27 June 2024).
8. Fisher, K. A Critical Psychosocial Spatiality of Learning. Doctoral Dissertation, Flinders University of South Australia, Bedford Park, Australia, 2002.
9. Sainsbury, L.; Long, R. *Health and the Environment: A Compilation of Evidence*; Australian Institute of Health and Welfare: Darlinghurst, Australia, 2011. Available online: https://www.aihw.gov.au/reports/environment-and-health/health-environment-compilation-of-evidence/summary (accessed on 27 June 2024).
10. Giles-Corti, B.; Badland, H.; Mavoa, S. Creating Healthy Liveable Neighbourhoods. T. A. P. P. Centre. 2017. Available online: https://preventioncentre.org.au/our-work/research-projects/creating-liveable-and-healthy-communities/ (accessed on 27 June 2024).
11. Well, I. International WELL Building Institute. Retrieved 27 February 2024. Available online: https://www.wellcertified.com (accessed on 27 June 2024).
12. Mazzi, A. Toward a Unified Language (and Application) of Salutogenic Design: An Opinion Paper. *Cent. Health Des.* **2021**, *14*, 337–349. [CrossRef]
13. Hughes, H.; Franz, J.; Willis, J. *School Spaces for Student Wellbeing and Learning: Insights from Research and Practice*; Springer Nature: Berlin/Heidelberg, Germany, 2019.
14. Clements-Croome, D. Flourish Theory: A model for multisensory human-centric design. In *A Handbook of Theories on Designing Alignment between People and the Office Environment*; Appel-Meulenbroek, R.D.V., Ed.; Routledge: London, UK, 2021; p. 12. [CrossRef]
15. Browning, W.; Ryan, C.; Clancy, J. *14 Patterns of Biophilic Design: Improving Health & Well-Being in the Built Environment*; Terrapin Bright Green: New York, NY, USA, 2014. Available online: https://www.terrapinbrightgreen.com/wp-content/uploads/2014/09/14-Patterns-of-Biophilic-Design-Terrapin-2014p.pdf (accessed on 27 June 2024).
16. Lei, Q.; Lau, S.; Yuan, C.; Qi, Y. Post-Occupancy Evaluation of the Biophilic Design in the Workplace for Health and Wellbeing. *Buildings* **2022**, *12*, 417. [CrossRef]
17. Alker, J. *Health, Wellbeing & Productivity in Offices: The Next Chapter for Green Building*; World Green Building Council: Ontario, CA, USA, 2024.
18. Lunny, C.; Kanji, S.; Thabet, P.; Haidich, A.; Boudioukas, K.; Pieper, D. Assessing the methodological quality and risk of bias of systematic reviews: Primer for authors of overviews of systematic reviews. *BMJ Med.* **2024**, *3*, e000604. [CrossRef]
19. Meyer, M.; Booker, J. Eliciting and analysing expert judgement: A practical guide. In *Society for Industrial & Applied Mathematics*; SIAM: Philadelphia, PA, USA, 2001.
20. Gillis, K.; Gatersleben, B. A Review of Psychological Literature on the Health and Wellbeing Benefits of Biophilic Design. *Buildings* **2015**, *5*, 948–963. [CrossRef]
21. Engemann, K.; Pedersen, C.B.; Arge, L.; Tsirogiannis, C.; Mortensen, P.; Svenning, J. Residential green space in childhood is associated with lower risk of psychiatric disorders from adolescence into adulthood. *Proc. Natl. Acad. Sci. USA* **2018**, *116*, 5188–5193. [CrossRef]
22. Hall, C.; Knuth, M. An Update of the Literature Supporting the Well-Being Benefits of Plants: A Review of the Emotional and Mental Health Benefits of Plants. *J. Environ. Hortic.* **2019**, *37*, 30–38. [CrossRef]
23. Kaplan, R.; Kaplan, S. *The Experience of Nature: A Psychological Perspective [Attention Restoration Theory (ART)]*; Cambridge University Press: Cambridge, UK, 1989.
24. Ulrich, R.; Simons, R.; Losito, B.; Fiorito, E.; Miles, M.; Zelson, M. Stress recovery during exposure to natural and urban environments. *J. Environ. Psychol.* **1991**, *11*, 201–230. [CrossRef]
25. Hodson, C.; Sander, H. Green urban landscapes and school-level academic performance. *Landsc. Urban Plan.* **2017**, *160*, 16–27. [CrossRef]
26. van den Bogerd, N.; Dijkstraa, S.; Koole, S.; Seidell, J.; de Vries, R.; Maas, J. Nature in the indoor and outdoor study environment and secondary and tertiary education students' well-being, academic outcomes, and possible mediating pathways: A systematic review with recommendations for science and practice. *Health Place* **2020**, *66*, 18. [CrossRef]
27. Vella-Brodrick, D.; Gilowska, K. Effects of Nature (Greenspace) on Cognitive Functioning in School Children and Adolescents: A Systematic Review. *Educ. Psychol. Rev.* **2022**, *34*, 1217–1254. [CrossRef]
28. Adams, S.; Savahl, S. Nature as children's space: A systematic review. *J. Environ. Educ.* **2017**, *48*, 291–321. [CrossRef]
29. Arola, T.; Aulake, M.; Ott, A.; Londholm, M.; Kouvonen, P.; Virtanen, P.; Paloniemi, R. The impacts of nature connectedness on children's well-being: Systematic literature review. *J. Environ. Psychol.* **2023**, *85*, 14. [CrossRef]

30. Davis, Z.; Guhn, M.; Jarvis, I.; Jerret, M.; Lorien, N.; Oberlander, T.; Sbihi, H.; Su, J.; van den Bosch, M. The association between natural environments and childhood mental health and development: A systematic review and assessment of different exposure measurements. *Int. J. Hyg. Environ. Health* **2021**, *235*, 113767. [CrossRef]
31. Johnstone, A.; Martin, A.; Cordovil, R.; Fjørtoft, I.; Iivonen, S.; Jidovtseff, B.; Lopes, F.; Reilly, J.; Thomson, H.; Wells, V.; et al. Nature-Based Early Childhood Education and Children's Social, Emotional and Cognitive Development: A Mixed-Methods Systematic Review. *Int. J. Environ. Res. Public Health* **2022**, *19*, 5967. [CrossRef]
32. Liu, J.; Green, R. The effect of exposure to nature on children's psychological well-being: A systematic review of the literature. *Urban For. Urban Green.* **2023**, *81*, 127846. [CrossRef]
33. Luo, W.; Kramer, R.; deKort, Y.; Rense, P.; Adam, J.; Lichtenbelt, W. Personal comfort systems and cognitive performance: Effects on subjective measures, cognitive performance, and heart rate measures. *Energy Build.* **2023**, *278*, 112617. [CrossRef]
34. Meng, X.; Zhang, M.; Wang, M. Effects of school indoor visual environment on children's health outcomes: A systematic review. *Health Place* **2023**, *83*, 103021. [CrossRef]
35. Russo, A.; Andreucci, M. Raising Healthy Children: Promoting the Multiple Benefits of Green Open Spaces through Biophilic Design. *Sustainability* **2023**, *15*, 1982. [CrossRef]
36. Robertson, J.; Martin, P.; Borradaile, L.; Alker, S. *Glasgow and the Clyde Valley Forest Kindergarten Feasibility Study;* Creative STAR Learning: Inverurie, Scotland, 2009. Available online: https://creativestarlearning.co.uk/wp-content/uploads/2020/07/Forest-Kindergarten-Report-April-2009.pdf (accessed on 27 June 2024).
37. Weiss, T.; Kahn, P.; Lam, L. Children's interactions with relatively wild nature associated with more relational behaviour: A model of child-nature interaction in a forest preschool. *J. Environ. Psychol.* **2023**, *86*, 101941. [CrossRef]
38. Collins, R.; Spake, R.; Brown, K.; Ogutu, B.; Smith, D.; Eigenbrod, F. A systematic map of research exploring the effect of greenspace on mental T health. *Landsc. Urban Plan.* **2020**, *201*, 1038230. [CrossRef]
39. Coventry, P.A.; Brown, J.; Pervin, J.; Brabyn, S.; Pateman, R.; Breedvelt, J.; Gilbody, S.; Stancliffe, R.; McEachan, R.; White, P. Nature-based outdoor activities for mental and physical health: Systematic review and meta-analysis. *SSM-Popul. Health* **2021**, *16*, 100934. [CrossRef]
40. Fernandez, A.; Ubalde-Lopez, M.; Yang, T.; McEachern, R.; Rashid, R.; Maitre, L.; Niewenhuisjen, M.; Vrijheid, M. School-Based Interventions to Support Healthy Indoor and Outdoor Environments for Children: A Systematic Review. *Environ. Res. Public Health* **2023**, *20*, 1746. [CrossRef]
41. Requiia, W.; Adams, M. Green areas and students' academic performance in the Federal District, Brazil: An assessment of three greenness metrics. *Environ. Res.* **2022**, *211*, 113027. [CrossRef]
42. Assem, H.; Khodeir, L.; Fathy, F. Designing for human wellbeing: The integration of neuroarchitecture in design—A systematic review. *Ain Shams Eng. J.* **2023**, *14*, 102102. [CrossRef]
43. Bower, I.; Tucker, R.; Enticott, P. Impact of built environment design on emotion measured via neurophysiological correlates and subjective indicators: A systematic review. *J. Environ. Psychol.* **2019**, *66*, 101344. [CrossRef]
44. Coburn, A.; Vartanian, O.; Kenett, Y.; Nadal, M.; Hartung, F.; Hayn-Leischsenring, G.; Navarrete, G.; Gonzalez-Mora Chatterjee, A. Psychological and neural responses to architectural interiors. *Cortex* **2020**, *126*, 217–241. [CrossRef]
45. Karakas, T.; Ylidiz, D. Exploring the influence of the built environment on human experience through a neuroscience approach: A systematic review. *Front. Archit. Res.* **2020**, *9*, 236–247. [CrossRef]
46. Luo, W.; Kramer, R.; Kompier, M.; Smolders, K.; de Kort, Y.; van Marken Lichtenbelt, W. Personal control of correlated color temperature of light: Effects on thermal comfort, visual comfort, and cognitive performance. *Build. Environ.* **2023**, *238*, 110380. [CrossRef]
47. Parhizkar, H.; Taddei, P.; Weziak-Bialowolska, D.; McNeely, E.; Spengler, J.; Laurent, J. Objective indoor air quality parameters and their association to respiratory health and well-being among office workers. *Build. Environ.* **2023**, *246*, 110984. [CrossRef]
48. Joye, J. Architectural lessons from environmental psychology: The case of biophilic architecture. *Rev. Gen. Psychol.* **2007**, *11*, 305–328. [CrossRef]
49. Ulrich, R. Aesthetic and affective response to natural environment. In *Human Behavior and Environment;* Springer: Boston, MA, USA, 1983. [CrossRef]
50. Valentine, C. The impact of architectural form on physiological stress: A systematic review. *Front. Comput. Sci.* **2014**, *5*, 1237531. [CrossRef]
51. Abdelaal, M.; Soebarto, V. Biophilic campus: An emerging planning approach for a sustainable innovation-conducive university. *J. Clean. Prod.* **2019**, *215*, 1445–1456. [CrossRef]
52. Khair, N.; Ajis, A.; Sakip, S. The Influence of Biophilic Design to Learning Ability in Library Environment: A Systematic Review. In Proceedings of the Virtual Go Green Conference and Publication: Rethinking Built Environment: Towards a Sustainable Future, Virtual, 29–30 September 2021.
53. Peters, T.; D'Penna, K. Biophilic Design for Restorative University Learning Environments: A Critical Review of Literature and Design Recommendations. *Sustainability* **2020**, *12*, 7064. [CrossRef]
54. Saarani, P.; Yusof, N.; Azir, K.; Khalid, Z. Biophilic Design in Improving the Psychological Performance of University Students: A Systematic Literature Review (SLR). In Proceedings of the Virtual Go Green Conference & Publication 2021, "Rethinking Built Environment: Towards a Sustainable Future", Virtual, 29–30 September 2021.

55. van den Bogerd, N.; Dijkstra, S.; Koole, S.S.J.; Maas, J. Greening the room: A quasi-experimental study on the presence of potted plants in study rooms on mood, cognitive performance, and perceived environmental quality among university students. *J. Environ. Psychol.* **2021**, *73*, 11. [CrossRef]
56. Yilmaz, S.; Ayten, A. A research on biophilic design patterns: The case of AGU as a biophilic campus. *Int. J. Archit. Plan.* **2023**, *11*, 692–715. [CrossRef]
57. Yassein, G.; Abrahiem, S. Biophilic Design in the Built Environment to Improve Well-Being: A Systematic Review of Practices. *J. Urban Res.* **2018**, *30*, 19. [CrossRef]
58. Maslow, A. A theory of human motivation. *Psychol. Rev.* **1943**, *50*, 370–396. [CrossRef]
59. Cripps, K. *Too Hot, Too Cold, and Then Just Right: Balancing Form and Function in Biophilic Designed Office Workspaces*; University of Maryland: College Park, MD, USA, 2019.
60. Zhang, Z.; Ye, B.; Yang, W.; Gao, Y. Effect of Nature Space on Enhancing Humans' Health and Well-Being: An Integrative Narrative Review. *Forests* **2024**, *15*, 100. [CrossRef]
61. Gascon, M.; Triguero-Mas, M.; Martinez, D.; Dadvand, P.; Rojas-Rueda, D.; Plasencia, A.; Nieuwenhuisjsen, M. Residential green spaces and mortality: A systematic review. *Environ. Int.* **2016**, *86*, 60–67. [CrossRef]
62. Gonçalves, G.; Sousa, C.; Fernandes, M.; Almeida, N.; Sousa, A. Restorative Effects of Biophilic Workplace and Nature Exposure during Working Time: A Systematic Review. *Int. J. Environ. Res. Public Health* **2023**, *20*, 6986. [CrossRef] [PubMed]
63. Sadick, A.; Kamarden, I. Enhancing employees' performance and well—Being with nature exposure. *J. Build. Eng.* **2020**, *32*, 101789. [CrossRef]
64. Tekin, B.; Corcoran, R.; Gutierrez, R. A Systematic Review and Conceptual Framework of Biophilic Design Parameters in Clinical Environments. *Cent. Health Des.* **2023**, *16*, 18. [CrossRef] [PubMed]
65. Tekin, B. *Human-Centred Therapeutic Environments: A New Framework for Biophilic Design*; University of Liverpool: Liverpool, UK, 2023.
66. LeFosse, D.; Timmeren, A.; Ratti, C. Biophilia Upscaling: A Systematic Literature Review Based on a Three-Metric Approach. *Sustainability* **2023**, *15*, 15702. [CrossRef]
67. Oxford Hub. Nature Positive: A Definition. 2024. Available online: https://naturepositive.web.ox.ac.uk/nature-positive-oxford (accessed on 27 June 2024).
68. Wijesooriya, N.; Bramgilla, A. Bridging biophilic design and environmentally sustainable design: A critical review. *J. Clean. Prod.* **2020**, *283*, 124591. [CrossRef]

architecture

MDPI

Article

Re-Thinking Biophilic Design for Primary Schools: Exploring Children's Preferences

Rokhshid Ghaziani

School of Architecture, Art and Design, University of Portsmouth, Portsmouth PO1 2DJ, UK; rokhshid.ghaziani@port.ac.uk

Abstract: Schools can play a vital role in supporting children's health and well-being, considering the time they spend in these environments. There is an increasing recognition of the role of nature in school design as many studies acknowledge that children have a critical need to be connected to nature. Therefore, the application of biophilia can be a design resolution appropriate for schools because of its impact on children's health and well-being; however, it remains relatively unexplored in school design around the world, including in the UK. Biophilic design patterns can be used in school buildings and grounds for greater connectivity between spaces and nature. This study focuses on biophilic design patterns related to direct and indirect experience of nature in the school environment. This paper discusses the findings of an empirical study with 88 children in a primary school and the application of biophilic design features and patterns related to happiness for primary school design. Finally, this paper suggests that the evaluation of biophilic design patterns in schools could be a co-design process with children and young people, essential for creating a healthy and happy school environment.

Keywords: biophilic school design; nature connectedness in primary schools; children's happiness; children's well-being; connecting schools to nature

1. Introduction

Children's lives today are very different from those of a few generations ago, with fewer opportunities to connect with nature. Urbanisation, which has meant the loss of green spaces, limiting their play outdoors, has played a huge role in this detachment. Furthermore, advancements in technology have encouraged children to spend more time indoors on electronic devices [1]. Studies show that the increased amount of time children spend disconnected from nature causes concern for their health and psychological well-being. The lack of interaction with nature leads to obesity, depression and loneliness [2]. It is reflected in Richard Louv's book *Last Child in the Woods* that children need to be saved from the "wired generation", described as having "nature deficit disorder" and favouring an indoor lifestyle, which leads to feeling detached from nature and being susceptible to stress, the loss of attention and having a negative mindset [2].

School environments play a significant role in influencing students' mental health and well-being [3]. A study by the NHS found that approximately one in five children and young people (20.3% of 8- to 16-year-olds) had a probable mental disorder in 2023 [4]. Therefore, the consideration of designing schools for well-being has become increasingly significant, considering the recent statistics; however, many school designs are characterised by spaces with inadequate fresh air, natural light and views of the outdoors, leading to a disconnection from the natural environment [5].

It has been highlighted that the natural environment plays an important role in maintaining and enhancing mental health and well-being [6,7]. Studies show that time in nature—direct and indirect contact, and engaging with nature through simple activities—is beneficial to well-being [8,9]. As children spend one-third of their time at school, the need for the reintegration of nature into schools is more important than ever [10]. There has been a rise in research that advocates that nature needs to be integrated into schools through biophilic design, as children are born "biophilic beings", and, hence, there is a need to learn about and explore nature through their built environment [11]. Kellert argues that nonetheless, values of nature should be brought back into the built environment using biophilic design, due to individuals being disconnected to nature as technology has developed and no longer perceiving nature as a necessity [12]. It is through this that children now have mindsets that nature is "nice" to have rather than a necessity.

However, as Louv states, this "nature deficit disorder" will lead to stunted academic and developmental growth [2]. Interactions with nature are fundamental to children's development, as they are instinctively driven by their curiosity about the natural environment. Therefore, nature is an important contributor to their learning and growth [13]. Children should understand that nature co-exists with humans and is a relationship that should be regarded as a high priority [13]. This nature connectedness is also important for children, providing essential sensory experiences that are instrumental to their physical and mental development [14].

At urban schools, students who encounter natural environments in their break times return to class with increased sustained and selective attention relative to their counterparts in built-up environments [15]. Studies show that among children aged 10 to 11 years in England [16] and 11 to 12 years in Australia [17], those who expressed more connection to nature reported a greater sense of well-being. Moreover, children ages 7 to 11 years in England who visited nature reserves for activities (from a day to more than 6 weeks) simultaneously reported increased nature connectedness, health and well-being [18].

Research into the biophilic design of schools is fairly new and limited. However, there have been a number of studies. Ghaziani et al. [19] discussed ten biophilic design patterns for schools related to two categories of nature in space and natural analogues. Watchman et al. [20] formulated a visual biophilic design vocabulary for schools. Browning and Determan [21] assessed the outcomes of biophilic design in elementary schools, and Aminpour [22] found that primary children attending Australian vertical schools have an affinity for biophilic elements. Abdul Malek et al. [23] conducted a comparative study on biophilic performances in elementary schools in Asia. However, these preferences have not been explored in primary schools in the UK.

This paper aims to identify the potential role of biophilic design features to promote children's happiness from their perspective. The empirical study examines children's preferences for features associated with eight patterns of biophilia related to direct and indirect connection to nature. It discusses what the most preferable biophilic features that need to be properly incorporated into the design process of primary schools are. Finally, this study provides recommendations and proposes further research on the application of biophilic features for the design of primary schools to support children's well-being.

2. Background

Nowadays, more and more children have less and less contact with the natural world, and this is having a huge impact on their health and development [24]. Richard Louv calls it "nature deficit disorder". The standard educational setting prioritises classroom-based abstract learning, which removes students from their connection to nature. However, as mentioned in the book *Schools that Heal*, one of the design strategies that support students' mental health and well-being is providing nature-filled environments, which improve mental and physical health [25]. Well-being in school environments is repeatedly approached as remedial; however, it could be approached as enabling young people to achieve their potential, promoting resilience, creativity and independence. Schools are not just places to fill young people with knowledge as proved during the pandemic, school design for well-being is important [26]. Grigoriou [27], in the book *Wellbeing in Interiors*, highlights the impacts of different features related to biophilic design patterns on well-being, including lighting, natural materials, colour, air quality and flow.

It has been also argued that well-being is intrinsically linked to happiness, suggesting that well-being consists of "fitting happiness". This theory asserts that happiness is characterised by a positive balance of affective states, including emotions and sensory pleasures. Therefore, well-being is defined through fitting happiness, establishing a nuanced relationship between the two concepts [28]. Higher levels of happiness correlate with improved physical and mental health. Various concepts and measurements of well-being highlight that happiness contributes to health through neurobiological processes, health behaviours, psychosocial resources and stress buffering effects. Overall, happiness is a significant determinant of well-being, impacting health outcomes [29].

Research on biophilic design patterns integrated into schools has gained recognition in recent years because studies show how children learn better when they interact with nature. School gardens offer significant benefits with regard to learning through experience [30]. Moreover, a number of design attributes related to the natural environment have been studied with regard to their educational impacts. A 14.4% improvement in test scores was discovered to result from natural ventilation [31]. The combination of dynamic lighting and an increased ventilation rate was observed to have a positive impact on the speed and concentration of children [32]. Furthermore, improved outdoor space and access to nature was linked to a 7% improvement in test results [33]. In addition, experiences with nature promote children's academic learning and development [34].

In addition, as discussed in the book *The Third Teacher*, most of the outdoor spaces designed for children fail to delight children. It is important to naturalise play spaces, allowing grass and leafy plants to flourish. This provides endless opportunities for play and discovery, as children want outdoor areas filled with nature, from plants, trees, flowers and water to animals and insects. Moreover, biophilic design is linked to sensory design. The childhood environment, including schools, constitutes an enormous workshop of the senses, which are an integral part of learning—an active element with the well-chosen application of colour, light, sound and smell in spaces. Children are a laboratory for senses, with each sense activating other senses. It is important to design schools that feed senses such as sound, smell, taste, touch, and movement and power memory, as an environment rich in sensory experiences helps students retain and retrieve what they learn [35]. While vision is the primary sense for most humans, and accounts for much of the brain's sensory capacity, the experience can be through scent, sound, touch, taste, temperature, pressure, balance, distance and more [36]; therefore, a sensual approach for the school design process is essential.

The application of biophilic design patterns [36] in school grounds and indoor spaces can create a stronger connection between interior design and nature that promotes children's well-being. Kellert [37], in his *Kinship to Mastery*, observed a progression in the bonding process with nature. While pre-schoolers focus on attraction, desire, fear and aversion experiences with nature, children between the ages 6 and 9 transition to a more emotional relationship with nature, and at about 10 years old and onwards, the relationship with nature becomes increasingly intellectual.

As Browning and Ryan [38] discuss, biophilia as a design philosophy does not lend itself well to quantification and is primarily a qualitative framework with endless solutions depending on the needs and priorities of people and places. Many green building standards and rating systems, such as BREEAM (a sustainability rating system for the performance of the built environment) and the WELL Building Standard (a body that measures the success of design for well-being and health) incorporate biophilic design. Biophilic design combines different components that connect directly to promote wellness and preserve the environmental sustainability of built facilities.

While sustainable buildings are well-established concepts in the construction industry and have the support of building regulations and certifications, the associated elements of building design aimed at promoting human well-being lack presence within the construction process and are often seen to be forfeited for financial gain or the ease of planning consent. As discussed by Browning and Ryan [38], many elements of biophilic design overlap and address both well-being and environmental concerns within the built environment.

Primary schools that adopt biophilic design patterns could make essential improvements in children's wellness and achievement levels and their physical and mental health. For example, biophilic classroom design has been shown to reduce stress, enhance creativity and clarity of thought and increase awareness and happiness in young people [39]. According to Determan et al. [39], there is a need for studies to investigate the impact of the integration of biophilic design patterns on different cognitive, psychological and academic outcomes in different educational environments. Future school designs and renovations require gathering children's views and collaboration among headteachers, teachers, architects and policymakers to implement nature-based school design strategies that will create enriched environments that allow children to develop optimally through their connection with nature.

3. Methods

To explore children's views related to the biophilic features of indoor and outdoor spaces in their school, a study was conducted in a primary school in Coventry, West Midlands, UK. Earlsdon Primary School, which opened in 1890, has been situated in the southwest of the city and is housed in a Victorian building. Figure 1 presents the school building and outdoor spaces, while Figure 2 shows indoor spaces.

Figure 1. Outdoor: school building (**a**) and the school playground (**b**).

In order to obtain permission for this research, two steps were carried out: (1) gaining authorisation from the Ethics committee at the Faculty of Art, Design and Humanities, De-Montfort University and (2) obtaining permission from those under the schools' authority—headteachers and parents (carers)—by having them review the information sheets and complete the consent forms. Children between the ages of 8 and 11 participated in this study. The printed questionnaires were distributed by the schoolteachers (to those children that had permission to participate in this study and were interested) and were completed during school hours.

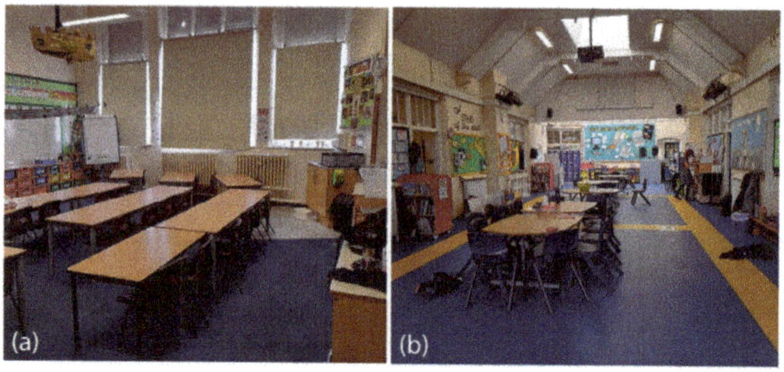

Figure 2. Indoor spaces: classroom (**a**), multipurpose space (**b**).

This study uses a tool previously established to gather voices in primary schools (ages 7–11) by Ghaziani et al. [19] and to include the features related to various biophilic design patterns (under the first two categories of Nature in Space (direct experience) and Natural Analogues (indirect experience)). The recommended features are associated with biophilic design patterns (based on the literature review and case studies) to select the common patterns—eight out of ten, as presented in Table 1 [19].

Table 1. Features associated with selected biophilic design patterns [19].

Theme	No.	Patterns	Features
Nature in the Space (Direct Experience)	1	Visual Connection with Nature	- Animals (e.g., birds and pets) - Landscape in school ground - Plants inside the classrooms
	2	Non-Visual Connection with Nature	- Sound of water - Sound of birdsong - Smell of flowers - Natural materials to touch (bamboo, wood and stone)
	3	Non-Rhythmic Sensory Stimuli	None
	4	Thermal and Airflow Variability	- A lot of fresh air from the windows
	5	Presence of Water	- A pond in school ground - An aquarium in the building
	6	Dynamic and Diffuse Light	- Lots of natural light from the windows - Skylight/roof window (in classrooms and school hall)
	7	Connection with Natural Systems	- View to outside to see plants and trees - Plants to grow and look after
Natural Analogues (Indirect Experience)	8	Biomorphic Forms and Patterns	- Natural form for seats and spaces - Circular or oval windows - Patterns of plants on walls (flowers, leaves) - Patterns on creatures on walls and floors (butterflies, shells) - Curved forms and spaces - Images of landscape on walls - Images of seaside on walls
	9	Material Connection with Nature	-Natural materials (bamboo and wood) inside the building to see and touch - Natural materials in school ground (bamboo, woodand stone) - Colourful walls and ceiling - Colourful glasses on the windows and doors
	10	Complexity and Order	None

The collection of ordinal data was chosen for this study. According to Bryman [40], ordinal data are based on counts of items assigned to specific categories that stand in some clear, ordered and ranked relationship. Therefore, it allows finding the importance of the identified items through the use of rating scales. The four-point scoring scale specifies one as the poorest score and four as the best score, as follows:

1. Sad;
2. I do not mind;
3. Happy;
4. Very happy.

In order to assess children's happiness in relation to diverse biophilic features, an evaluative tool was applied [19]. The questionnaire included an image for each biophilic feature (beside each written item) in order to help children visualise the items and share their views, as Figure 3 illustrates.

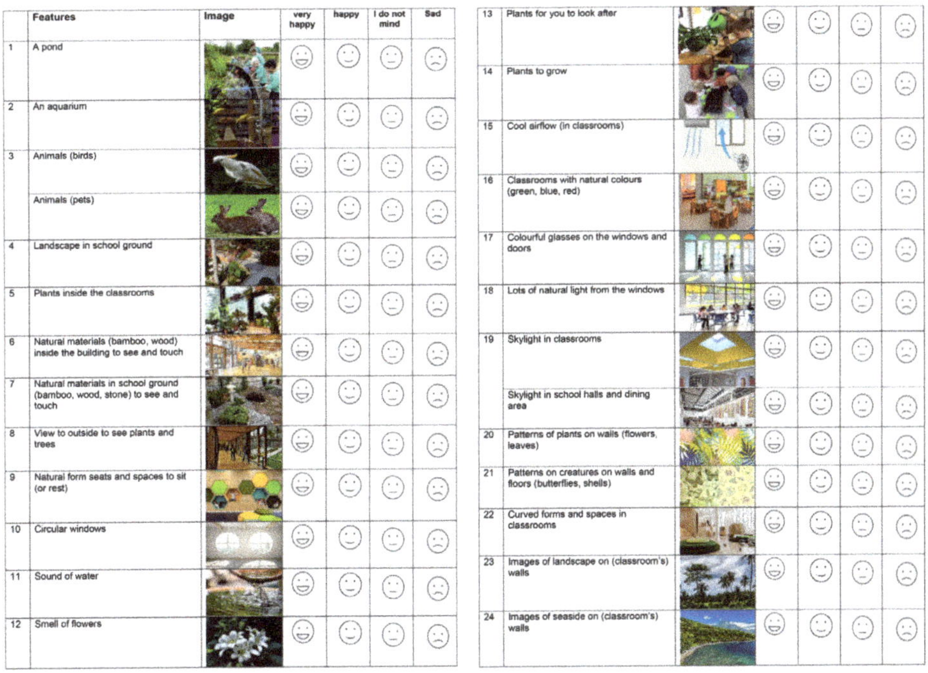

Figure 3. Format of questionnaire.

The question for children (written above the table with images) was "What do you feel about having these features in your school? (Please choose one of the faces for each feature (draw a circle around it or add a tick under it)". A total of 88 children (44 Year 4 and 44 Year 6) participated in this study; however, an analysis of gender could not be made because some children did not write their gender (boy or girl) on the questionnaires.

4. Findings

As Figure 4 presents, among the biophilic features, pets (79.3%) and aquariums (71.3%) were chosen to contribute the most to children's happiness. Similarly, views of plants and trees (69.4% very happy) as well as natural form seating (70.9% very happy) were associated with happiness. These findings indicate that natural elements and nature-informed features can provide feelings of peace and comfort.

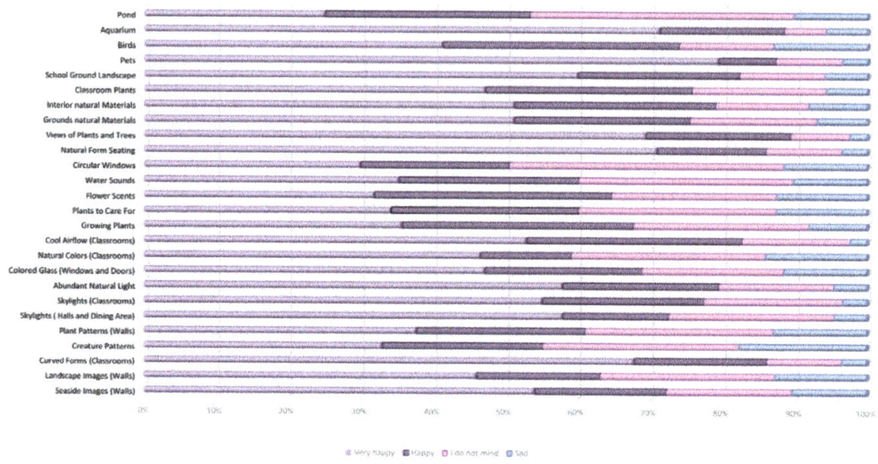

Figure 4. Children's responses related to various biophilic features.

Other features, such as natural light (57.9% very happy) and skylights in classrooms and communal areas (55.9–57.9% very happy) contribute to children's feelings of happiness. In addition, classrooms with cool airflow (52.9% very happy) have a positive impact on children's happiness.

Some of the features were considered neutral, including ponds (36.4% neutral) and circular windows (37.9% neutral). While these features may contribute to an aesthetically pleasing environment, they may not be sufficient to affect children's happiness.

In order to find the overall happiness related to various biophilic features, the responses to "happy" and "very happy," were merged. Figure 5 presents the positive impact on children's happiness. "Views of plants and trees" had the highest happiness at 89.4%, followed very closely by "aquariums" at 88.5% happiness. It is also shown that "pets" have a strong positive effect on 87.4% of students. Furthermore, "natural form seating" (86%) and "cool airflow in classrooms" (82.8%) were considered highly impactful for children's happiness.

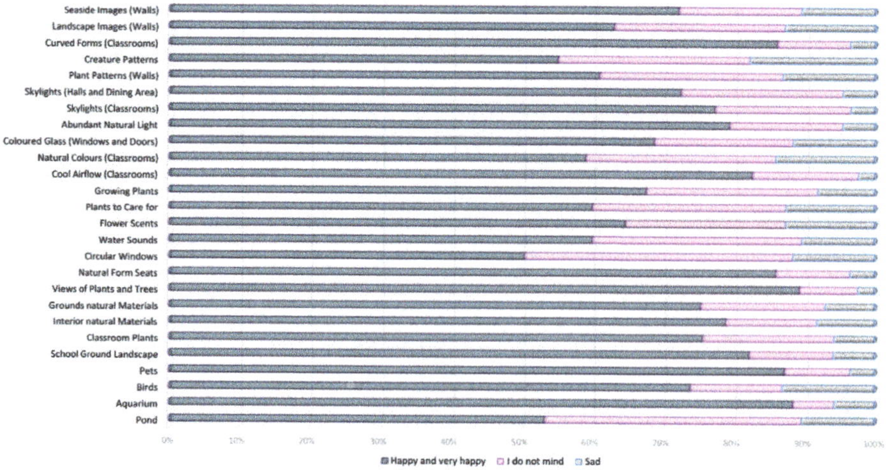

Figure 5. Overall happiness related to biophilic features.

5. Discussion

The biophilic features can be discussed in three categories based on children's overall responses to happiness, as follows:

(1) The most preferred features (65% and above) were "aquarium" (77%), having "pets" at the school (76%), the opportunity of "viewing plants and trees" (76%), "natural form seating" (74%) and the presence of "school ground landscape" features (70%). "Interior natural materials" (such as wood, bamboo and stone to see and touch) were rated as 68%, while the placement of "plants in classrooms" received a rating of 66%, followed by school grounds' natural materials (65%). These features are related to "Presence of Water" (pattern No. 5), "Visual Connection with Nature" (pattern No. 1), "Connection with Natural System" (pattern No. 7), "Biomorphic Forms and Pattern" (pattern No. 8) and "Material Connection with Nature" (pattern No. 9).

(2) Medium preferable biophilic features in relation to happiness (55–64%) were "the scent of flowers" (63%) followed by "plants to care for" (59%) and "water sound" (57%). These features are related to "Non-Visual Connection with Nature" (pattern No. 2) and "Connection with Natural System" (pattern No. 7).

(3) The least preferred features (below 54%), were "pond" (53%) and "circular windows" (50%), indicating limited impact on children's happiness. These features are related to "Presence of Water" (pattern No. 5) and "Biomorphic Forms and Pattern" (pattern No. 8).

The findings show that the majority of biophilic features could have positive impacts on primary school children's happiness; however, the wider impacts could include stress reduction, cognitive performance benefits and emotional well-being benefits (see Ghaziani et al. [19]). As happiness and well-being are closely linked, considering the biophilic design of schools to promote children's happiness is important. The blending of biophilic design in educational settings has also been shown to reduce stress indicators like blood pressure and heart rate, as well as improving learning experiences. Educational environments that connect students to nature enable producing spaces that merge academic achievements with student wellness benefits [39].

Moreover, by implementing biophilic design, schools can efficiently deliver nature-based learning by using nature as an educational resource for direct hands-on experiences essential for forming their environmental opinions. This approach not only fosters a deeper connection between students and their environment but also instils a sense of responsibility towards sustainable practices that they can carry into their future endeavours. Besides, from an ecological point of view, incorporating biophilic design in schools could cultivate appreciation and inspire children to become involved with efforts to protect the environment and the habitats of wildlife.

It is also important to consider the link between biophilic design and building standards. In addition to the well-being (including happiness) aspect of biophilic school design, its incorporation in rigorous standards such as BREEAM (Building Research Establishment Environmental Assessment Method), along with the WELL Building Standard, which prioritises the health and well-being of building occupants, can significantly enhance and elevate the overall educational environment in which students learn and grow [38]. BREEAM addresses biophilic design primarily in the Health & Wellbeing (HEA) section of the rating system (HEA 01: Visual Comfort, HEA 07: Safe and Healthy Surroundings), although opportunities to gain credit through particular biophilic design patterns can also be found in Land Use & Ecology (LE 04: Ecological Change and Enhancement).

However, with the WELL Building Standard™ (WELL), biophilic design is primarily addressed in the Mind section (Mind 02: Access to Nature). It requires that indoor environments include direct connection to nature through views or indoor natural features (water, plants and light), indirect connection to nature through natural materials or images and spatial layouts that use environmental psychology to enhance experience (as well as Mind 07: Restorative Space). Moreover, several categories, including Air (Air 07: Operable Windows), Light (Light 05: Enhanced Daylight Access, Light 03: Circadian Light Design), Thermal Comfort (03: Thermal Zoning) and Sound (05: Sound Masking), grant opportunities to use particular biophilic design patterns [38].

The significant impact of biophilic patterns on children's emotional responses allows us to (re)design various spaces in schools based on the importance of biophilic patterns related to children's age and preferences. According to the findings of this study (for children between the ages of 8 and 11), architects and designers should consider the implications of the most preferred biophilic features in primary school design, including "aquarium", "pet corners" and providing views of "plants and trees" in different spaces, including classrooms. Moreover, "landscape" in school grounds and "plants in classrooms" need to be considered, as do "natural form for seats" and "natural materials" in interior spaces. Finally, it is crucial to consider the interdisciplinary aspect of biophilic design in schools. Educators and psychologists, along with architects and designers, should create learning spaces that foster both educational success and mental wellness (happiness and satisfaction) for school children.

6. Conclusions

This paper identifies the potential role of biophilic design features in promoting children's happiness in a primary school from their own perspectives. The empirical study explored children's preferences related to eight patterns from two categories of direct and indirect connection to nature. It discussed the most preferred biophilic features in relation to children's happiness that need to be properly incorporated into the design process of primary schools. The findings show the majority of biophilic features could have positive impacts on primary school children's happiness. They indicate that both direct and indirect connection to nature in a primary school are important to children's happiness, suggesting that biophilic design principles are crucial in a school environment. As happiness and well-being are closely linked, with research indicating that higher levels of happiness correlate with improved physical and mental health, considering the biophilic design of schools to promote children's happiness is a key determinant of successful school design.

In addition, this study has its limitations, as the data was collected in one primary school in the UK and the analysis could not relate the responses (children's preferences) to their gender; therefore, more research needs to be carried out in various primary (elementary) and secondary schools to find the most effective and age-appropriate biophilic design features to be considered in the school design process. Involving children and young people as experts in a collaborative biophilic design process, essential for creating a healthy and happy school environment, is recommended. Finally, considering an interdisciplinary approach that includes architects, educators, psychologists, public health professionals, ecologists and regenerative designers would be beneficial for the better design of schools in connection to nature.

Funding: This research received no external funding.

Institutional Review Board Statement: The study was conducted in accordance with the Declaration of Helsinki, and approved by the Institutional Review Board of Faculty of Arts, Design and Humanities, De Montfort University (G32 and 10 December 2020).

Informed Consent Statement: Informed consent was obtained from all subjects involved in the study.

Data Availability Statement: The data could be available on request from the corresponding author.

Conflicts of Interest: The author declares no conflict of interest.

References

1. Kellert, S.R.; Heerwagen, J.H.; Mador, M.L. *Biophilic Design: The Theory, Science, and Practice of Bringing Buildings to Life*; Wiley: Hoboken, NJ, USA, 2008.
2. Louv, R. *Last Child in the Woods: Saving Our Children from Nature-Deficit Disorder*; Algonquin Books of Chapel Hill: Chicago, IL, USA, 2005.
3. Higgins, S.E.; Hall, E.; Wall, K.; Woolner, P. *The Impact of School Environments: A Literature Review*; University of Newcastle: Newcastle, UK, 2005. Available online: https://eprints.ncl.ac.uk/12574 (accessed on 10 April 2025).
4. NHS. One in Five Children and Young People Had a Probable Mental Disorder in 2023. 2023. Available online: https://www.england.nhs.uk/2023/11/one-in-five-children-and-young-people-had-a-probable-mental-disorder-in-2023/ (accessed on 20 April 2025).
5. Largo-Wight, E. Cultivating healthy places and communities: Evidenced-based nature contact recommendations. *Int. J. Environ. Health Res.* **2011**, *21*, 41–61. [CrossRef] [PubMed]
6. Capaldi, C.; Passmore, H.-A.; Nisbet, E.; Zelenski, J.; Dopko, R. Flourishing in nature: A review of the benefits of connecting with nature and its application as a wellbeing intervention. *Int. J. Wellbeing* **2015**, *5*, 1–16. [CrossRef]
7. Pritchard, A.; Richardson, M.; Sheffield, D.; McEwan, K. The Relationship between Nature Connectedness and Eudaimonic Well-Being: A Meta-analysis. *J. Happiness Stud.* **2019**, *21*, 1145–1167. [CrossRef]
8. Bowler, D.E.; Buyung-Ali, L.M.; Knight, T.M.; Pullin, A.S. A systematic review of evidence for the added benefits to health of exposure to natural environments. *BMC Public Health* **2010**, *10*, 456. [CrossRef]
9. McMahan, E.A.; Estes, D. The effect of contact with natural environments on positive and negative affect: A meta-analysis. *J. Posit. Psychol.* **2015**, *10*, 507–519. [CrossRef]
10. Kellert, S. *Three. The Practice of Biophilic Design*; Yale University Press eBooks: New Haven, CT, USA, 2019; pp. 23–110. [CrossRef]
11. Kellert, S. Dimensions, Elements, and Attributes of Biophilic Design. In *Biophilic Design*; John Wiley Sons Inc.: Hoboken, NJ, USA, 2008; pp. 3–20.
12. Kellert, S. *Building for Life: Designing and Understanding the Human-Nature Connection*; Island Press: Washington, DC, USA, 2005.
13. Park, S.J.; Lee, H.C. Spatial design of childcare facilities based on biophilic design patterns. *Sustainability* **2019**, *11*, 2851. [CrossRef]
14. Arola, T.; Aulake, M.; Ott, A.; Lindholm, M.; Kouvonen, P.; Virtanen, P.; Paloniemi, R. The impacts of nature connectedness on children's well-being: Systematic literature review. *J. Environ. Psychol.* **2022**, *85*, 101913. [CrossRef]
15. Amicone, G.; Petruccelli, I.; De Dominicis, S.; Gherardini, A.; Costantino, V.; Perucchini, P.; Bonaiuto, M. Green Breaks: The restorative effect of the school environment's green areas on children's cognitive performance. *Front. Psychol.* **2018**, *9*, 01579. [CrossRef]
16. Richardson, M.; Sheffield, D.; Harvey, C.; Petronzi, D. *The Impact of Children's Connection to Nature: A Report for the Royal Society for the Protection of Birds (RSPB)*; RSPB: Derby, UK, 2016. [CrossRef]
17. Whitten, T.; Stevens, R.; Ructtinger, L.; Tzoumakis, S.; Green, M.J.; Laurens, K.R.; Holbrook, A.; Carr, V.J. Connection to the natural Environment and Well-Being in Middle Childhood. *Ecopsychology* **2018**, *10*, 270–279. [CrossRef]
18. Sheldrake, R.; Amos, R.; Reiss, M. *Children and Nature: A Research Evaluation for the Wildlife Trusts*; The Wildlife Trusts: Newark, UK, 2019; Available online: https://discovery.ucl.ac.uk/id/eprint/10087410/ (accessed on 10 April 2025).
19. Ghaziani, R.; Lemon, M.; Atmodiwirjo, P. Biophilic design patterns for primary schools. *Sustainability* **2021**, *13*, 12207. [CrossRef]
20. Watchman, M.; DeKay, M.; Demers, C.M.H.; Potvin, A. Design vocabulary and schemas for biophilic experiences in cold climate schools. *Archit. Sci. Rev.* **2021**, *65*, 101–119. [CrossRef]
21. Browning, W.; Determan, J. Outcomes of Biophilic Design for Schools. *Architecture* **2024**, *4*, 479–492. [CrossRef]
22. Aminpour, F. Child-friendly environments in vertical schools: A qualitative study from the child's perspective. *Build. Environ.* **2023**, *242*, 110503. [CrossRef]
23. Malek, N.A.; Atmodiwairjo, P.; Wungpatcharapon, S.; Ghaziani, R. A comparative study on biophilic preferences of school learning settings: A case of elementary schools in Asia. *Plan. Malays.* **2024**, *22*, 544–563. [CrossRef]
24. Henley, J. Why Our Children Need to Get Outside and Engage with Nature. *The Guardian*. 2010. Available online: https://www.theguardian.com/lifeandstyle/2010/aug/16/childre-nature-outside-play-health (accessed on 12 April 2025).
25. Chawla, L. Schools That Heal: Design with Mental Health in Mind by Claire Latané. *Child. Youth Environ.* **2022**, *32*, 185–187. [CrossRef]
26. Gillen, N.; Nissen, P.; Park, J.; Scott, A.; Singha, S.; Taylor, H.; Taylor, I.; Featherstone, S. *Rethink Design Guide*; Taylor & Francis: London, UK, 2021. [CrossRef]
27. Grigoriou, E. *Wellbeing in Interiors*; Taylor & Francis: London, UK, 2019. [CrossRef]
28. Rossi, M.; Tappolet, C. *Well-Being as Fitting Happiness*; Oxford University Press eBooks: Oxford, UK, 2022; pp. 267–289. [CrossRef]

29. Mendes, M.; Simões, P.; Simões, J.A.; Santiago, L.M.; Prazeres, F.; Maricoto, T. The link between happiness and health: A review of concepts, pathways and strategies for enhancing well-being. *Fam. Med. Prim. Care Rev.* **2023**, *25*, 288–296. [CrossRef]
30. McCarty, J.; Ford, V.; Ludes, J. Growing Experiential learning for the future: REAL School gardens. *Child. Educ.* **2018**, *94*, 47–55. [CrossRef]
31. Shaughnessy, R.J.; Haverinen-Shaughnessy, U.; Nevalainen, A.; Moschandreas, D. A preliminary study on the association between ventilation rates in classrooms and student performance. *Indoor Air* **2006**, *16*, 465–468. [CrossRef]
32. Hviid, C.A.; Pedersen, C.; Dabelsteen, K.H. A Field Study of the Individual and Combined Effect of Ventilation Rate and Lighting Conditions on Pupils' Performance. *Build. Environ.* **2020**, *171*, 106608. [CrossRef]
33. Tanner, C. The Influence of School Architecture on Academic Achievement. *J. Educ. Adm.* **2000**, *38*, 309–330. [CrossRef]
34. Kuo, M.; Barnes, M.; Jordan, C. Do experiences with nature promote learning? converging evidence of a Cause-and-Effect relationship. *Front. Psychol.* **2019**, *10*, 305. [CrossRef] [PubMed]
35. O'Donnell Wicklund Pigozzi and Peterson; Architects Inc.; VS Furniture; Bruce Mau Design. *The Third Teacher: 79 Ways You Can Use Design to Transform Teaching & Learning*; Harry Abrams: New York, NY, USA, 2010.
36. Ryan, C.O.; Browning, W.D.; Clancy, J.O.; Andrews, S.L.; Kallianpurkar, N.B. BIOPHILIC DESIGN PATTERNS: Emerging Nature-Based Parameters for Health and Well-Being in the Built Environment. *Int. J. Archit. Res. Archnet-IJAR* **2014**, *8*, 62. [CrossRef]
37. Kellert, S.R. *Kinship to Mastery: Biophilia in Human Evolution and Development*; Island Press: Washington, DC, USA, 1997.
38. Browning, W.D.; Ryan, C.O. *Nature Inside*; RIBA Publishing: London, UK, 2020. [CrossRef]
39. Determan, J.; Akers, M.A.; Albright, T.; Browning, B.; Martin-Dunlop, C.; Archibald, P.; Caruolo, V.; Macht, C. Impact of Biophilic Learning Spaces on Student Success. 2019. Available online: https://cgdarch.com/wp-content/uploads/2019/12/The-Impact-of-Biophilic-Learning-Spaces-on-Student-Success.pdf (accessed on 16 April 2025).
40. Bryman, A. *Quantity and Quality in Social Research*; Oxford University Press: London, UK, 2003. [CrossRef]

architecture

MDPI

Article

Designing Thriving School Ecosystems: The Synergy of Biophilic Design, Wellbeing Science, and Systems Science

Fiona Gray [1],* and Andrea Downie [2],*

1 Independent Researcher, Victoria 3331, Australia
2 Centre for Wellbeing Science, Melbourne Graduate School of Education, University of Melbourne, Melbourne 3010, Australia
* Correspondence: fiona.gray@biolivingbydesign.com.au (F.G.); andrea@projectthrive.com.au (A.D.)

Abstract: This article presents a novel approach that integrates biophilic design, wellbeing science, and systems science into a holistic strategy designated as the Biophilic Wellbeing Systems Approach (BWSA). This transdisciplinary approach aims to enhance student wellbeing in educational settings by fostering a deep connection with nature, supported by positive psychology and systems thinking. The research explores how these combined disciplines can shape educational environments that improve learning outcomes and promote human thriving. The study advocates for a transformative redesign of school environments, emphasising the interdependence of individual and planetary health, drawing on Indigenous wisdom and contemporary scientific knowledge. This foundational strategy holds significant potential to fundamentally change how schools are designed and operated. By fostering environments that deeply integrate wellbeing principles, this approach not only influences the physical buildings but also extends to the entire school ecosystem. It works synergistically to promote educational excellence and holistic wellbeing, ensuring that all elements of the school environment contribute to the overall development and health of students and staff.

Keywords: biophilic design; systems thinking; wellbeing science; education; positive psychology; environmental design; educational architecture; sustainability; Eudaimonic education; indigenous wisdom

1. Introduction

The importance of wellbeing in educational settings has increasingly been recognised over the last few decades, leading to an increase in research, funding, programs, and initiatives aimed at enhancing students' mental and physical health [1]. Despite these efforts, the role of environmental design in promoting wellbeing has often been overlooked. School environments play a significant role in influencing students' mental health and wellbeing [2,3], and while there is a robust and growing body of evidence supporting the benefits of well-designed, nature-connected educational spaces [4,5], many schools persist in designing buildings that fail to incorporate elements and strategies that deliver these benefits.

Australian students receive a total of 11,000 h of instruction in a classroom throughout primary and high school, the highest in the OECD. Furthermore, they spend around 74 percent more time in the classroom than their Finnish counterparts but achieve significantly lower scores in international assessments compared to students in similar countries [6]. This discrepancy indicates that increased classroom time does not necessarily equate to academic success, suggesting the need for a more comprehensive approach that considers the overall wellbeing of students to enhance educational outcomes.

Wellbeing is a multidimensional construct with different meanings across ages, cultures, and population groups [7]. The most widely cited academic definition of wellbeing describes it as a measure of how people feel and function on a personal and social level and how they evaluate their lives as a whole [7]. Concepts like meaning, values, personal growth, and happiness are often used to describe wellbeing. While wellbeing is widely recognised as critically important, strategies to enhance human and planetary health often lack an interdisciplinary approach. In education systems, efforts are often compartmentalised, with wellbeing siloed into specific programs, subjects, or frameworks. Additionally, wellbeing is frequently overlooked as an architectural imperative, missing opportunities to design spaces that actively promote students' mental and physical health. Traditional focuses on cost-efficiency, utilitarian functionality, and conventional building norms prioritise immediate practical concerns over the long-term health benefits of well-designed spaces.

Long-established theories such as the Attention Restoration Theory [8] and Stress Reduction Theory [9] underscore the psychological benefits of exposure to nature, including improved cognitive function and reduced stress levels. These foundational theories have been reinforced by more recent research demonstrating similar benefits [10,11]. The concept of Biophilia, initially articulated by Wilson [12] and further developed by Kellert [13], posits an innate human affinity for nature that can be harnessed to enhance learning environments. Contemporary studies continue to support this idea, showing that integrating natural elements into educational settings can lead to significant improvements in student wellbeing and academic performance [14,15]. These theories and findings provide a robust foundation for understanding the positive impacts of biophilic design on wellbeing in educational contexts.

The issue of designing for wellbeing in educational settings has become increasingly critical in light of recent statistics. The latest Mission Australia Youth Survey [16], which collected responses from 19,501 participants, revealed that 30% of respondents identified mental health as a significant issue in Australia today, with 32% expressing that they were personally 'extremely' or 'very concerned' about mental health. Environmental issues topped the list of concerns for the second consecutive year at 40%. Globally, mental health challenges among young people are notably high, with one in seven individuals aged 10–19 years experiencing a mental disorder, accounting for 13% of the global disease burden in this age group [17]. In Australia, mental ill-health affects nearly half of all adults over their lifetimes, with suicide being the leading cause of death among Australians aged 15 to 49 years in 2019 [18]. These statistics underscore the urgent need for innovative approaches that integrate wellbeing into educational environments.

Concurrently, the loss of natural habitats, biodiversity, and the impacts of climate change pose serious risks to both human and planetary health. These challenges highlight the urgent need for sustainable practices across all sectors, including education and construction. Biophilic design, an 'approach that emphasises the necessity of maintaining, enhancing and restoring the beneficial experience of nature in the built environment' [13], offers a promising solution. By integrating biophilic design into schools, there is an opportunity not only to enhance student and staff wellbeing but also to promote environmental stewardship, contributing to a healthier planet.

2. Research Method and Contribution

This article contends that a systems thinking approach is essential for recognising the profound impact of the built environment on wellbeing in order to foster thriving educational ecosystems. It introduces the Biophilic Wellbeing Systems Approach (BWSA), a novel synthesis of biophilic design, wellbeing science, and systems science applied to educational settings. Applying a grounded theory methodology, the research adopts an exploratory approach to generate insights that inform a new theoretical approach [19]. Starting with a broad interdisciplinary lens rather than a specific research problem, the study allows the theory to emerge inductively from the research, incorporating multiple perspectives and emphasising the interactions and interconnectedness across various fields.

Data collection involved an exploratory review of the literature across the disciplines of biophilic design, wellbeing science, and systems science, as they relate to education and school environments, to identify intersections, overlaps, and potential synergies. This was complemented by in-depth interviews with architects and school leaders, along with analyses of buildings at pioneering case-study schools in Australia and Indonesia. Through the qualitative analysis of the collected data and synthesis of knowledge from these different fields, the study aims to develop a novel, transdisciplinary framework. This approach transcends traditional disciplinary boundaries by integrating insights and perspectives from these diverse fields, demonstrating how this integration can enhance educational environments and student wellbeing.

The discussion critiques traditional educational structures, arguing that they often fail to meet the developmental needs of contemporary learners. Applying an architectural lens and incorporating principles of positive psychology and Eudaimonic Education® (a registered trademark of Project Thrive Australia Pty Ltd., Highton, Australia), this approach proposes a transformative method to create school environments that foster educational achievement and genuine thriving.

This study delves into the integration of biophilic design, wellbeing science, and systems science, recognising that wellbeing emerges not simply from direct teaching but is also cultivated—'caught', 'taught', and 'sought' [20]—through the thoughtful design of environments, learning experiences, and cultural contexts. These elements foster wellbeing by enabling a reconnection with nature, incorporating both ancient knowledge and modern science to sustain human and planetary health.

The article also explores the alignment of Indigenous wisdom and systems thinking, examining how these ways of knowing can contribute to creating holistic learning environments. Case studies were selected from public, independent, and private schools to illustrate the practical implementation and benefits of these concepts. The discussion culminates with practical applications and recommendations for future research and implementation, emphasising the significant potential of the BWSA in educational settings.

By establishing a foundation for future research and practical applications, this study proposes an innovative strategy for educational settings. It makes an original contribution by being the first to propose and explore this transdisciplinary approach, offering a unique perspective that bridges multiple fields to enhance student wellbeing and learning outcomes in a highly integrated manner.

3. Biophilic Design, Indigenous Wisdom, and Systems Thinking

The evolution of human society is deeply intertwined with nature, a relationship embedded in the concept of biophilia, which denotes an innate human affinity with the natural world. This bond is crucial for both our psychological and physiological wellbeing and has been a cornerstone of human evolution [21]. This connection is not just a contemporary observation but is rooted in ancient wisdom, where life and environment were understood as holistic and indivisible.

Indigenous societies worldwide have historically exemplified a seamless integration with nature, often reflected in their sustainable practices, spiritual beliefs, and land management techniques that emphasise coexistence with the natural environment. Central to many Indigenous worldviews is the belief in the interconnectedness of all life, vividly illustrated through rich narratives and cultural practices. The Creation stories of Aboriginal Australians are one such example that highlight a symbiotic relationship with the environment that is deeply embedded in their social structures, rituals, and daily practices [22].

Indigenous cultures have long embraced what is now recognised through a Western lens as systems thinking [23]. A large body of work within systems science explicitly acknowledges and addresses the complexity of the real world, including interrelationships of its elements, multiple perspectives, temporality (i.e., changes over time) and the challenges this creates in replicating results, shifting patterns of behaviour, unintended consequences, and recognition that parts cannot be studied in isolation from the whole [24]. Systems science does not view environmental elements in isolation but as parts of a comprehensive system that affects and supports all aspects of life. Such perspectives harmonise with biophilic design principles that aim to create environments supportive of life.

Integrating Indigenous knowledge and perspectives into modern educational frameworks provides a powerful means to enhance holistic learning experiences and, consequently, overall wellbeing. This integration aligns educational practices with systems thinking, which acknowledges and leverages the complex interactions between nature, cultural heritage, and environmental stewardship. A fundamental principle of biophilic design is the emphasis on place-based relationships [25] (p. 12) [26] which enables educational environments to reflect the geographical, historical, ecological, and cultural identities of their settings. Such an approach not only enriches the learning experience but also fosters a deep, enduring connection to the local context, thereby cultivating a sense of identity and belonging among students.

In this light, integrating Indigenous knowledge and biophilic design principles in education is more than a stylistic or cultural inclusion; it signifies a fundamental rethinking of how learning environments can function as interconnected ecosystems that honour and reflect the timeless human connection to nature. This approach offers a meaningful way to create nurturing, sustainable spaces that are deeply rooted in their local contexts and responsive to the environmental challenges of our time.

4. Biophilic Design and Systems Thinking in Educational Environments

Biophilic design and systems thinking hold the potential to profoundly transform school environments by deepening the connections between learning spaces and the broader contexts of the natural world, societal dynamics, and educational frameworks. Viewing educational environments through this holistic lens acknowledges the complex interplay among various elements and recognises humans as integral components of a broader ecosystem. In harmony with Indigenous wisdom, it appreciates the interconnectedness inherent in nature and society. More than just enhancing aesthetic appeal, incorporating natural elements into educational settings creates dynamic, living 'learning laboratories' where direct engagement with nature can cultivate a deeper understanding of ecological principles and instil a strong sense of environmental responsibility in students.

Embracing systems thinking enables students to see beyond isolated facts, aiding their understanding of the complexity and interdependence of natural systems. This insight is particularly crucial in today's volatile, uncertain, complex, and ambiguous (VUCA) world. It equips learners with the skills to identify the root causes of issues, anticipate future challenges, and formulate effective strategies [27]. Such a perspective fosters an appreciation of how various elements within an ecosystem interact, highlighting the need for comprehensive solutions that respect and address these complex relationships.

5. Positive Psychology and Wellbeing Science

Significantly contributing to our global understanding of wellbeing has been the rapidly growing field of Positive Psychology (PP), often referred to as wellbeing science. PP studies the conditions that contribute to the optimal functioning of people, groups and institutions, and aims to improve quality of life [28]. However, as noted by researchers and practitioners like Kern et al. [24] in 2019, the hype regarding the field's potential has at times led to exaggerated claims, over-inflated expectations, disillusionment, dismissal, and unintentional harms. It has also been observed that PP interventions and programs sometimes overlook the complex and dynamic realities and varied contexts in which people reside [24]. To address these issues, Systems Informed Positive Psychology (SIPP) was proposed as an evolution of PP. This approach explicitly incorporates principles and concepts from the systems sciences into positive psychology theory, methodologies, practices, and discourse, to optimise human social systems and the individuals within them [24]. This perspective acknowledges that human wellbeing depends on more than just the individual and programs. The SIPP approach aims to optimise human social systems and the individuals within them, facilitating system co-evolution and the envisioning of positive, unimagined futures [24].

SIPP has been effectively integrated into educational contexts, most recently through practical work focused on Systems and Eudaimonic Education (SEE) [29] which involves creating conditions that support meaning, purpose, authenticity, and personal growth through environmental, cultural, and learning design, grounded in wellbeing science. This approach emphasises the importance of designing educational environments that not only support academic achievement but also enhance overall wellbeing. The success of these holistic design principles in promoting comprehensive wellbeing has led to the development of the 'wholebeing' framework, specifically tailored for educational settings.

Jarden and Downie [30] define 'wholebeing' as an emergent phenomenon resulting from high levels of wellbeing, low levels of illbeing, and the presence of resilience. This framework enables individuals to express their true selves and lead a sustainable, fulfilling life characterised by authenticity, value-driven actions, and a focus on both self and others, optimising their performance in a learning environment. They further suggest that higher levels of wholebeing correlate with better physical health, and vice versa.

The Wholebeing Model (Figure 1) incorporates a systems thinking approach and is supported by factors that enhance personal growth and performance. It emphasises the individual's relationships with others, their environment, and the resources available to them, including time, finances, and energy. The model considers the interconnected relationships across different systemic levels—Me, We, and Us—illustrating how individuals, groups, and larger communities thrive together.

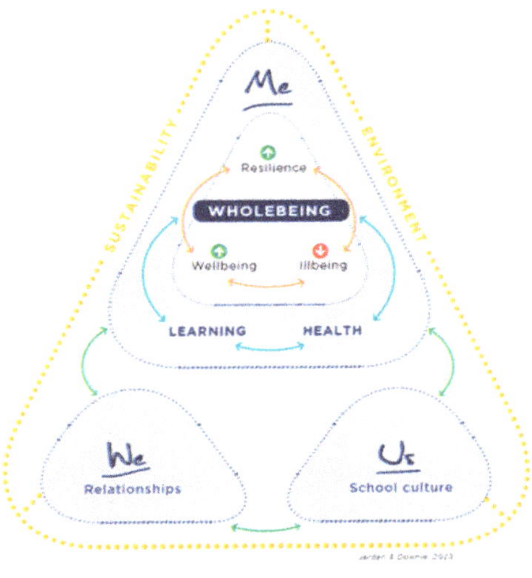

Figure 1. The Wholebeing Model. Source: [30].

The Wholebeing Model assists systems leaders in comprehending the broad, interconnected system, which is crucial for developing a shared understanding of complex challenges [31] while recognising possibilities and opportunities. Jarden and Downie [30] argue that this is because it fosters reflection, directs resource utilisation, facilitates generative conversations, and shifts the focus from merely reacting to problems related to illbeing towards co-creating joint responsibility for a better future incorporating all relevant factors. The environment encompasses the entire model, encapsulating its relationships with all other core components while underscoring the environment's critical role in enhancing wellbeing, reducing mental ill-health, and cultivating adaptability and resilience. This comprehensive influence illustrates the necessity and significance of the built environment in educational settings.

By applying biophilic design principles to enhance the environmental element of the Wholebeing Model, a new, holistic approach to the design of school environments is introduced. While synergies have been identified between biophilic design principles and systems science, as well as between wellbeing science/positive psychology and systems science [24,25], the integration of all three disciplines—biophilic design, systems science, and positive psychology—has not yet been explored. The BWSA incorporates each of these elements in the educational and architectural design of schools to create environments that promote nature connection, support learning, and bolster wellbeing.

6. Biophilic Design in Schools

In the realm of educational architecture, the paucity of biophilic elements within typical school environments warrants further attention and improvement. Many educational settings are characterised by enclosed spaces that lack adequate natural light, fresh air, and views of the outdoors, leading to a disconnection from the natural environment [32]. This issue is particularly critical given the substantial amount of time students spend within these environments. Additionally, empirical studies indicate that static and uninspired school environments, often characterised by monochromatic lighting, white walls, and confined layouts, can negatively impact students' cognitive function [33,34]. Such detachment severs the essential connection with nature and adversely affects key aspects of students' psychological and physiological health.

Engagement with nature, whether through direct contact or simulated elements, taps into our deep-seated evolutionary instincts and is crucial for designing educational spaces that fulfill our innate needs for safety, exploration, and harmonious coexistence with our environment. This connection to nature is especially vital for children, as it provides essential sensory experiences that are instrumental in their physical and mental development [35].

For instance, buildings that harness the sun's movement to create dynamic light and shadow patterns can foster a connection to the natural rhythm of days and seasons, aligning with our internal biorhythms to produce restorative effects [36]. This alignment offers a counterbalance to the fast pace of contemporary life, while also enriching learning experiences. Natural elements such as water and plants can also play a significant role [26]. Water features like small ponds or fountains, and the integration of plant life in school settings, have a calming effect, reducing stress and anxiety. The sound and sight of water, combined with the visual appeal of greenery, can lower blood pressure and enhance concentration, making these elements not just aesthetically pleasing but also conducive to students' overall health and wellbeing. Incorporating varied natural features also enriches the sensory experience, further fostering a connection with nature and enhancing the educational environment.

In situations where direct access to natural elements is limited, simulating natural features within educational environments can still offer significant benefits [37]. By incorporating elements that mimic the aesthetics and ambience of nature, such as natural shapes, colours, sounds, and materials, schools can partially compensate for the absence of direct natural exposure. Tapping into our innate preference for natural environments, the inclusion of such elements can provide psychological and physiological benefits typically associated with direct contact with nature.

7. A Paradigm Shift in Educational Architecture, Pedagogy, and Wellbeing Design

The principles of biophilic design are not just theoretical; they are being actively implemented around the world, signalling a paradigm shift in educational architecture and pedagogy. This shift encompasses changes in teaching practices, curriculum design, approaches to wellbeing, and the physical spaces designed for learning. As schools embrace these concepts, they not only change their physical structures but also the very nature of how education is delivered and experienced. The tangible impact of this holistic approach is illustrated in the following three case studies, which detail the practical implementation and beneficial outcomes in diverse educational settings and sectors (public, private, and independent) in Australia and Indonesia.

8. Green School, Bali

The Green School in Bali, Indonesia, conceptualised by John and Cynthia Hardy, serves as an exemplar of biophilic educational architecture. It adeptly integrates the natural environment in a manner that is both innovative and reflective of Indigenous vernacular traditions. The school prioritises ecological building materials and techniques, while concurrently promoting the wellbeing of its students by focusing on social and emotional learning, supported by systems thinking tools within its curriculum.

Central to the architectural ethos of the Green School is the innovative use of bamboo, a material chosen for its sustainability, rapid growth rate, strength, and adaptability. The extreme bend in the school's bamboo structures serves as a visual representation of the fundamental wellbeing principle and life skill of 'adaptability' which has been shown to be a greater predictor of satisfaction with life, self-esteem, school participation, and a sense of meaning and purpose than resilience [38]. In its use throughout the school, bamboo has been transformed from a traditional material for temporary structures to the primary component of permanent, robust, and elegant structures across the campus. These include the Heart of the School, the Millennium Bridge that connects the east and west sides of the campus, and the school's latest architectural addition, The Arc. Each of these structures showcases bamboo's versatility in creating vast, open communal spaces designed to foster learning and social interaction.

In the Heart of the School, towering bamboo columns support multi-level spaces, blending sophisticated engineering with local craftsmanship (Figure 2). Even the bamboo flooring, constructed using bamboo pins without synthetic adhesives or finishes, reflects the school's dedication to environmental integrity and creating a healthy learning environment for children. The building is deliberately positioned to create a strong sense of connection and belonging. The notion of connection is further highlighted in the architectural expression by the structure's three interconnected spiralling nautilus forms. These fluid, organic forms break away from traditional, rectilinear education spaces, to create a physical environment that inspires creativity [39] and dynamic learning experiences. Classrooms, designed as open-air pavilions on rice fields, also maximise connectivity and immersion in nature, and each learning space is consistently furnished with bamboo, sustaining the biophilic theme.

The Arc, built in 2021, demonstrates an advanced application of biophilic shapes, forms, and patterns (Figure 3). Designed by the architecture studio Ibuku, led by Elora Hardy, and working in collaboration with carpentry specialist Jörg Stamm and Atelier One structural engineers, the building features bamboo arches spanning 19 m, interconnected by anticlastic gridshells. These gridshells derive strength from curving in opposite directions, enhancing the building's structural integrity. The architecture exemplifies the concept of biomimicry, emulating a mammal's ribcage stabilised by tensile membranes that function like tendons and muscles, efficiently distributing structural forces. Serving dual purposes as a wellness space and a gymnasium, The Arc enhances the campus facilities with a design that deeply integrates biological concepts in both form and function.

In its extensive use of bamboo, the school's architecture embodies flexibility, strength, and lightness—qualities that mirror human adaptability and wellbeing, as well as resonate with the Balinese philosophy of Tri Hita Karana. This philosophy promotes a symbiotic relationship between people, nature, and the spiritual world, principles that the Balinese consider essential for human thriving, and which are seamlessly integrated into the school's core educational values. By embodying these ideals, the architectural design reinforces the school's commitment to a holistic and interconnected learning approach.

The Green School's pioneering approach has gained considerable acclaim, leading to the planning and establishment of three additional international campuses and its acquisition by the educational conglomerate Education in Motion (EiM). Through its innovative practices, the Green School serves as an illustration of how architecture can respect and enhance the relationship between the built environment and the natural world while also influencing wellbeing and educational outcomes. As a revolutionary shift in educational design, the Green School in Bali has inspired educators around the world (Figure 4) and directly influenced educational projects in other regions, including the following case study from Australia.

Figure 2. Towering bamboo columns supporting multi-level spaces of the Heart of the School. Photo: Fiona Gray.

Figure 3. The rib-like structure of The Arc, designed by Ibuku (2021). Photo: Fiona Gray.

Figure 4. School principals participating in a principals redesigning education immersion program led by Project Thrive (2019). Photo: Andrea Downie.

9. Lisieux Catholic Primary School, Australia

Lisieux Catholic Primary School in Torquay, Australia, represents a progressive model in educational innovation, blending systems thinking and biophilic design with a deep respect for cultural heritage. Under the leadership of founding principal, Dr. Susan Ryan, the school was established with a clear vision. This vision was further shaped by a collaboration with the second author (Downie), the director of Project Thrive—a consulting firm that specialises in education redesign, wellbeing, and systems leadership. A pivotal experience in this collaboration was an educational redesign immersion in Bali that included a visit to the Green School [40]. The school has dedicated itself to boosting wellbeing by merging Positive Psychology with thoughtful environmental design that supports connection with nature and committed respect for Indigenous cultural practices. The school's design establishes a tangible connection to nature and place, fostering a learning environment conducive to educational outcomes while promoting a deeper understanding of, and respect for, the natural world and Aboriginal culture. It achieves this by merging natural elements, place-based cultural contexts, and sustainable practices.

Beginning in 2018, the school's construction has progressed in stages. The second stage of the development, designed by Minx Architecture (Figure 5), introduced a deliberate emphasis on enhancing connection with nature. This design phase incorporated a colour palette inspired by local natural elements such as Waratah, Spotted Gum, Banksia, and seaspray (J. Briese, personal communication, 18 April 2024). The school's principal has observed the positive impact this has had on the learning environment, observing that the calming colours enhance concentration and stimulate creative and innovative thinking (S. Ryan, personal communication, 21 April 2024).

Figure 5. Lisieux Catholic Primary School (2022), Torquay, Australia. Photo: Rhiannon Slatter.

Biophilic and cultural integration extends through the naming of the teaching spaces after native species—Acacia, Waratah, Spotted Eucalypt, Golden Wattle, Grevillea, Scent-bark, and Silver Banksia—embedding local flora into the learning environment. Botanical motifs represented on decals of glazed internal partitions add another layer of natural imagery, strengthening the connection to the school's local context.

The school's engagement with Aboriginal culture is multifaceted, woven into both the curriculum and the physical environment. The celebration of Stage Two's opening with a traditional smoking ceremony, an ancient Aboriginal practice that involves smouldering native plants to produce smoke believed to hold cleansing properties and the ability to ward off bad spirits, underscores this deep respect (Figure 6). The school also collaborates with local Aboriginal Elders on special projects, such as the creation of a possum skin cloak by Aunty Lisa Couzens. This cloak, rich with stories of belonging, place, and the sacred and spiritual aspects of Aboriginal culture, is worn by Aboriginal students during assemblies, graduation ceremonies, and other significant events. Indigenous knowledge is further embedded in the student's learning through a unique Beach Discovery Program, where they engage with local beaches to learn about sustainability, environmental care, and marine life. This program offers an authentic, hands-on learning experience about connection to Country, cultural heritage sites, traditional ecological knowledge, respect, and reconciliation.

Nature is woven into the school's fabric through the naming of its four house teams after native birds: Bundjil, Kunuwarra, Gherang, and Parrwang. In collaboration with Aboriginal artist-in-residence Norm Jurrawaa Stanley, students helped design a playground mural that visually represents these birds, enhancing the visual landscape of the school and enhancing students' engagement with local fauna and Aboriginal heritage. Cultural connections are further emphasised in the school's upstairs breakout space, named the Wayapperi Hub, meaning "meeting space". This area serves multiple purposes and fosters interaction, enhancing the communal feel of the school environment.

Figure 6. Cultural education programs facilitator, Norm Jurrawaa Stanley, playing the didgeridoo at the opening ceremony of the stage two building at Lisieux Catholic Primary School. Photo: Susan Ryan.

The school has strategically incorporated design features to ensure natural light floods the communal spaces and is borrowed into the classrooms through generous skylights above a central double-height atrium. In many schools, deep floor plans often result in substantial distances from the building's exterior walls to its interior core, typically due to large classrooms connected by hallways or other interior spaces. This layout can make it challenging to bring daylight into central areas. However, access to daylight supports the body's natural circadian rhythm, which regulates cycles of alertness and sleepiness, and has been shown to enhance mood, increase alertness, and improve overall productivity and learning outcomes in educational settings [41]. By integrating natural light, the school creates a healthier and more engaging environment for students and staff.

Biophilic design at Lisieux also emphasises the integration of outdoor and natural spaces. Classrooms with access to outdoor areas, many featuring breakout balconies or direct access to outdoor spaces, support an immersive outdoor educational approach. This is exemplified by dedicated areas like the Outdoor Discovery space and the Ubuntu Garden, which facilitate direct contact with nature and hands-on learning experiences integrated into the curriculum.

Through these design choices, Lisieux illustrates a systems-based approach to education that prioritises sustainability, cultural integration, and a meaningful connection to the natural world, demonstrating how educational environments can function as ecosystems of learning and respect for heritage and nature.

10. Wollert Primary School, Australia

Wollert Primary School in Victoria, Australia, is a government school designed as a Supported Inclusion Hub, tailored to enhance learning outcomes for students with diverse abilities and needs. The architectural partnership of Thomson Adsett and Law Architects has brought to life a design that emphasises inclusion and accessibility, with specialised facilities like sensory areas, accessible play spaces, and classrooms designed with acoustic, visual, and tactile elements to support all students (Figure 7).

Figure 7. Wollert Primary School, Wollert, Victoria (2022). Photo: Dianna Snape.

A key biophilic principle integrated into the school's design is the concept of prospect and refuge, which balances spaces for expansive observation with areas that provide a sense of protection. Features such as large windows and bleacher seating offer prospect, while elements like dark colours and deep bay windows with seating nooks offer safe, cocooning spaces that embody the notion of refuge (Figure 8). Informed by research from the Turner Institute for Brain and Mental Health at Monash University, the use of dark green walls in certain classrooms was specifically chosen to evoke this sense of refuge, offering a calming, enclosed space that helps mitigate sensory overload. This strategic use of colour aims to create a more comforting and protective atmosphere for those with heightened sensitivities.

Figure 8. Bay window, bleacher seating, and dark carpets in Wollert Primary School learning space (2022). Photo: Dianna Snape.

Working closely with educational planner Lisa Horton, the school was organised into 'learning neighbourhoods', each comprising four classrooms designed to support a range of educational experiences and pedagogical approaches (S. Brincat, personal communication, 2 February 2024). These neighbourhoods are crafted for both collaborative and individual learning, emphasising flexibility and inclusion. Each classroom is uniquely tailored to support different teaching methods and learning preferences. At the heart of each neighbourhood is a central area where classrooms converge, with distinct segments that support varied sensory and learning experiences. This setup facilitates both group activities and quieter areas for focused work, allowing teachers to organise students in environments that best suit their educational needs, enhancing the overall learning experience by aligning the physical space with the pedagogical approach.

Attention to acoustics and tactility was a paramount consideration in the design, with the use of acoustic panelling and varied wall textures and materials such as brick and timber that also provide a tactile quality that aids navigation for vision-impaired students while also enhancing the sensory and wayfinding experience for all students.

The landscape design by Outlines Landscape Architecture is an integral element of the school's biophilic strategy. It replaces the conventional concrete apron that typically surrounds school buildings with rich indigenous plantings that blur the indoor–outdoor boundary, enhancing visual permeability. To further incorporate natural elements, the design team collaborated with a local ecologist to integrate invertebrates into the outdoor setting. Various deceased insects were sourced and encapsulated in resin within rocks, enriching the educational qualities of the landscape and fostering a connection with the local ecosystem. Historical elements are also interwoven into the landscape, with remnants of the site's recent agricultural past, such as old fences, serving as historical markers that deepen the school's connection to its local context.

Community feedback influenced the school's development, shaping it into a space that serves as an extension of home and a community hub, especially crucial in an area where new suburbs feature housing stock with limited connection to nature and a car-centric urban layout. The design preserves areas of cultural significance and prioritises accessible natural spaces and outdoor learning environments, enhancing educational outcomes and fostering a deeper sense of environmental awareness (S. Brincat, personal communication, 2 February 2024).

Overall, the school presents a notable shift in educational architecture by deliberately accommodating the diverse sensory and cognitive needs of all students through the application of biophilic principles to enhance their wellbeing. This approach not only fosters an inclusive and adaptable educational experience but also supports the development of a dynamic and responsive learning community.

11. A Battle for Relevance

Schools around the world are recognising the need for a more relevant education system and within that, more relevant learning spaces [42]. Recent pedagogical shifts reflect a growing emphasis on experiential and authentic learning [43]. This highlights a need to reconsider the design of learning environments to enable the conditions that support this type of educational philosophy. Biophilic design intersected with wellbeing and systems science can forge dynamic and engaging learning environments that ignite curiosity and promote exploration for experiential and authentic learning to emerge. Such environments allow for direct and continuous interaction with nature, playing a crucial role in the holistic development of students. At a pivotal moment in wellbeing science, it is understood that the learning environments we create can positively impact the mental health, wellbeing, and outcomes of the individuals that reside within them [44]. Integrating biophilic design into educational environments is an important strategy for nurturing connections with the natural world. This approach enhances the emotional and psychological development of students, equipping future generations to be deeply attuned to environmental sustainability and concerns. It holds the potential to develop well-rounded individuals who possess a profound connection to and understanding of themselves, others, and their environment.

12. Future Research and Implementation Opportunities

This novel approach presents multiple opportunities for future research and practical application in schools. Longitudinal studies are essential to track the long-term impacts of the BWSA on student wellbeing and educational outcomes. Such studies will provide robust empirical evidence over extended periods to validate the effectiveness of the BWSA in diverse school environments. Conducting additional case studies in varied geographical and socio-economic contexts will further understanding of the adaptability and effectiveness of the BWSA across different educational settings. This will help identify best practices and potential challenges, ensuring that the approach can be tailored to meet the needs of different communities.

Developing and implementing quantitative measures that encompass psychological, emotional, social, and academic performance dimensions will be crucial in providing a comprehensive understanding of how the BWSA impacts overall student wellbeing. Future studies could also include ethnographic surveys to observe user responses and behavioural patterns. This will help increase understanding of how specific design elements impact student wellbeing, the extent of these impacts, the ways in which students interact with the designed environment, as well as identify areas for improvement.

Integrating Indigenous knowledge, with a focus on how Indigenous perspectives can enrich biophilic design principles and contribute to holistic and culturally responsive educational environments, is another promising area for research. Documenting these integrations will provide valuable insights into creating inclusive and diverse learning spaces.

The holistic nature of the BWSA also necessitates the development of comprehensive policy guidelines and implementation frameworks for educational authorities and school administrators. These guidelines should be developed by interdisciplinary teams of researchers, educators, architects, and policymakers, to provide practical steps and considerations for incorporating nature-connected elements in school environments. This collaborative approach aligns with the overarching ethos of the BWSA and will ensure the strategies are well-rounded and feasible for implementation.

Supporting the implementation of the BWSA can also be achieved through professional development programs for educators, architects, and school leaders. These programs should focus on increasing awareness and understanding of the BWSA, emphasising the benefits of biophilic design, wellbeing science, and systems thinking. Training should provide practical guidance on effectively applying these principles in educational settings, ensuring that all stakeholders are equipped to foster environments that enhance student wellbeing and learning outcomes. By addressing these areas, future research and practical applications can build on the foundational insights provided by this study, leading to more effective and holistic educational environments that support the wellbeing and development of students.

13. Conclusions

This article explored the dynamic synergy between biophilic design, wellbeing science, and systems science as vital elements in fostering holistic wellbeing within educational settings. This interdisciplinary approach investigates how the integration of nature and its complex processes supports and enhances the learning experience by aligning with the principles of Eudaimonic Education®. This approach is about creating environments that improve educational outcomes while simultaneously supporting wellbeing and preparing young people for future challenges.

Throughout the discussion, the article addresses the critical need for a shift away from traditional school designs that often neglect the well-documented benefits of connection to nature. Despite significant investment in wellbeing programs, many young people still struggle, emphasising the need for a comprehensive, integrated approach that sees educational spaces as ecosystems where wellbeing is cultivated through thoughtful design and an understanding of the interconnectedness of all life.

Indigenous knowledge systems, with their deep-rooted understanding of these connections, alongside the principles of biophilic design, provide a rich source of inspiration. These systems illustrate that learning environments can and should reflect and respect the natural world, fostering spaces that are not just places of learning but also sources of wellbeing. The significant impact of a harmonious relationship with the natural world on psychological and physiological wellbeing underscores the importance of this approach.

Key case studies at Green School Bali, Lisieux Catholic Primary School and Wollert Primary School demonstrate these concepts in action. These schools have broken new ground by creating learning environments that meaningfully integrate nature, honour Indigenous practices and local contexts, and prioritise ecological and sustainable practices. By also incorporating principles of wellbeing science, these schools support both the physical and psychological health of their students.

Ultimately, a paradigm shift in the conception and realisation of educational environments is needed. This shift involves embracing a holistic view that integrates biophilic design, wellbeing science, and systems thinking, an approach termed the Biophilic Wellbeing Systems Approach (BWSA). This approach represents a step towards fostering a more environmentally conscious, balanced, adaptable, and 'well' generation. By broadening the scope of conventional educational models, this approach deepens connection to the environment and enhances awareness of students' own roles within the natural world. This shift is crucial for developing well-rounded individuals who are equipped to navigate the complexities of a rapidly changing world. By fostering environments that reflect these principles, schools can play a crucial role in promoting sustainable living and wellbeing, preparing students for academic success and a meaningful, authentic, and healthy life, connected to their world and each other.

Author Contributions: Conceptualisation, F.G. and A.D.; methodology, F.G. and A.D.; data collection, F.G.; data analysis, F.G.; writing—original draft preparation, F.G. and A.D.; writing—review and editing, F.G. and A.D. All authors have read and agreed to the published version of the manuscript.

Funding: This research received no funding.

Informed Consent Statement: Informed consent was obtained from all interviewees involved in the study.

Data Availability Statement: The original contributions presented in the study are included in the article. Further inquiries can be directed to the corresponding authors.

Acknowledgments: The authors express their gratitude to Nikki Stanley for granting permission to reference and include an image of Norm Jurrawaa Stanley. We also acknowledge and pay our respects to Norm Jurrawaa Stanley, whose wisdom and contributions have left an important legacy. Thank you to Jane Briese, Sally Brincat, and Susan Ryan for sharing valuable insights and information regarding the case study schools.

Conflicts of Interest: Andrea Downie is a director of Project Thrive. The authors declare that the research was conducted in the absence of any commercial or financial relationships that could be construed as a potential conflict of interest.

References

1. White, M.; Murray, A.; Seligman, M.P. *Evidence-Based Approaches in Positive Education: Implementing a Strategic Framework for Well-Being in Schools*; Springer: Berlin/Heidelberg, Germany, 2015. [CrossRef]
2. Higgins, S.; Hall, E.; Wall, K.; Woolner, P.; McCaughey, C. *The Impact of School Environments: A Literature Review*; The Centre for Learning and Teaching-School Education, Communication and Language Science, University of Newcastle: Callaghan, Australia, 2005.
3. Barrett, P.; Davies, F.; Zhang, Y.; Barrett, L. The impact of classroom design on pupils' learning: Final results of a holistic, multi-level analysis. *Build. Environ.* **2015**, *89*, 118–133. [CrossRef]
4. Alves, S.; Gulwadi, G.B.; Nilsson, P. An Exploration of How Biophilic Attributes on Campuses Might Support Student Connectedness to Nature, Others, and Self. *Front. Psychol.* **2021**, *12*, 793175. [CrossRef] [PubMed]

5. Mann, J.; Gray, T.; Truong, S.; Sahlberg, P.; Bentsen, P.; Passy, R.; Ho, S.; Ward, K.; Cowper, R. A systematic review protocol to identify the key benefits and efficacy of nature-based learning in outdoor educational settings. *Int. J. Environ. Res. Public Health* **2021**, *18*, 1199. [CrossRef]
6. Singhal, P. Australian Students Spend More Time in Class, Are still Outperformed in Tests. 2018. Available online: https://www.smh.com.au/education/australian-students-spend-more-time-in-class-are-still-outperformed-in-tests-20180727-p4zu34.html (accessed on 27 April 2024).
7. Jarden, A.; Roache, A. What Is Wellbeing? *Int. J. Environ. Res. Public Health* **2023**, *20*, 5006. [CrossRef] [PubMed]
8. Kaplan, R.; Kaplan, S. *The Experience of Nature: A Psychological Perspective*; Cambridge University Press: Cambridge, UK, 1989.
9. Ulrich, R.S. View through a window may influence recovery from surgery. *Science* **1984**, *224*, 420–421. [CrossRef] [PubMed]
10. Kweon, B.S.; Ulrich, R.S.; Walker, V.D.; Tassinary, L.G. Anger and stress: The role of landscape posters in an office setting. *Environ. Behav.* **2008**, *40*, 355–381. [CrossRef]
11. Berto, R. The role of nature in coping with psycho-physiological stress: A literature review on restorativeness. *Behav. Sci.* **2014**, *4*, 394–409. [CrossRef]
12. Wilson, E.O. *Biophilia*; Harvard University Press: Cambridge, MA, USA, 1984.
13. Kellert, S.R.; Heerwagen, J.H. Preface. In *Biophilic Design: The Theory, Science, and Practice of Bringing Buildings to Life*; Wiley: Hoboken, NJ, USA, 2008.
14. Determan, J.; Akers, M.; Albright, T.; Browning, B.; Martin-Dunlop, C.; Archibald, P.; Caruolo, V.; Macht, C. *Impact of Biophilic Learning Spaces on Student Success*; The American Institute of Architects: Washington, DC, USA, 2019. Available online: https://www.brikbase.org/sites/default/files/The%20Impact%20of%20Biophilic%20Learning%20Spaces%20on%20Student%20Success.pdf (accessed on 8 July 2024).
15. Peters, T.; D'Penna, K. Biophilic Design for Restorative University Learning Environments: A Critical Review of Literature and Design Recommendations. *Sustainability* **2020**, *12*, 7064. [CrossRef]
16. Mission Australia. Youth Survey. 2023. Available online: https://www.missionaustralia.com.au/publications/youth-survey (accessed on 27 April 2024).
17. World Health Organization. Mental Health of Adolescents. 2021. Available online: https://www.who.int/news-room/fact-sheets/detail/adolescent-mental-health#:~:text=Key%20facts,illness%20and%20disability%20among%20adolescents (accessed on 27 April 2024).
18. Australian Government Department of Health and Aged Care. Mental Health and Suicide Prevention. 2024. Available online: https://www.health.gov.au/topics/mental-health-and-suicide-prevention (accessed on 27 April 2024).
19. Bryant, A. The Grounded Theory Method. In *The Oxford Handbook of Qualitative Research*, 2nd ed.; Leavy, P., Ed.; Oxford Academic: Oxford, UK, 2020; pp. 167–199.
20. Curren, R.; Boniwell, I.; Ryan, R.M.; Oades, L.; Brighouse, H.; Unterhalter, E.; Kristjánsson, K.; de Ruyter, D.; Macleod, C.; Morris, I.; et al. Finding consensus on well-being in education. *Theory Res. Educ.* **2024**, *22*, 117–157. [CrossRef]
21. Barbiero, G.; Berto, R. Biophilia as Evolutionary Adaptation: An Onto-and Phylogenetic Framework for Biophilic Design. *Front. Psychol.* **2021**, *12*, 700709. [CrossRef] [PubMed]
22. Stanner, W.E.H. The Dreaming (1953). In *The Dreaming & Other Essays*; Black Inc. Agenda: Collingwood, VIC, Australia, 2011; pp. 57–72.
23. Goodchild, M. Relational systems thinking: That's how change is going to come, from our earth mother. *J. Aware.-Based Syst. Chang.* **2021**, *1*, 75–103. [CrossRef]
24. Kern, M.L.; Williams, P.; Spong, C.; Colla, R.; Sharma, K.; Downie, A.; Oades, L.G. Systems informed positive psychology. *J. Posit. Psychol.* **2020**, *15*, 705–715. [CrossRef]
25. Kellert, S.R. *Nature by Design: The Practice of Biophilic Design*; Yale University Press: New Haven, CT, USA, 2018.
26. Browning, W.; Ryan, C.; Clancy, J.; 14 Patterns of Biophilic Design. Terrapin Bright Green. 2014. Available online: https://www.terrapinbrightgreen.com/report/14-patterns/ (accessed on 10 July 2024).
27. Johansen, B.; Euchner, J. Navigating the VUCA world. *Res.-Technol. Manag.* **2013**, *56*, 10–15. [CrossRef]
28. Seligman, M.E.P.; Csikszentmihalyi, M. Positive psychology: An introduction. *Am. Psychol.* **2000**, *55*, 5–14. [CrossRef]
29. Downie, A.K. Could Redesigning Education End the Mental Health Crisis? A Paradigm Shift Towards a More Relevant System. Medium. 2023. Available online: https://medium.com/@andrea-downie/could-redesigning-education-end-the-mental-health-crisis-3c099ec869d3 (accessed on 12 March 2023).
30. Jarden, A.; Downie, A.; Finter, K.; Jarden, R. Next level flourishing in education: A case study of 'wholebeing'. In *Handbook of Positive School Psychology Interventions: Evidence-Based Practice for Promoting Youth Mental Health*; Arslan, G., Yıldırım, M., Eds.; Springer: Berlin/Heidelberg, Germany, 2023.
31. Senge, P.; Hamilton, H.; Kania, J. The dawn of system leadership. *Stanf. Soc. Innov. Rev.* **2015**, *13*, 27–33. [CrossRef]
32. Largo-Wight, E. Cultivating healthy places and communities: Evidence-based nature contact recommendations. *Int. J. Environ. Health Res.* **2011**, *21*, 41–61. [CrossRef] [PubMed]
33. Ming, A.; Jinghua, H.; Yoshihiro, S.; Tetsuo, K. Time-of-day-dependent Effects of Monochromatic Light Exposure on Human Cognitive Function. *J. Physiol. Anthropol.* **2009**, *28*, 217–223. [CrossRef]
34. Grube, K. Detrimental Effects of White Valued Walls in Classrooms. *Educ. Plan.* **2014**, *21*, 69–82.

35. Arola, T.; Aulake, M.; Ott, A.; Lindholm, M.; Kouvonen, P.; Virtanen, P.; Paloniemi, R. The impacts of nature connectedness on children's well-being: Systematic literature review. *J. Environ. Psychol.* **2023**, *85*, 101913. [CrossRef]
36. Unnikrishnan, G. *WELL V2: Evidence behind the Light Concept*; International WELL Building Institute: New York, NY, USA, 2021. Available online: https://f.hubspotusercontent40.net/hubfs/7039796/Evidence%20Box/evidence%20box-%20light%20final.pdf (accessed on 1 May 2024).
37. Pasca, L.; Carrus, G.; Loureiro, A.; Navarro, Ó.; Panno, A.; Tapia Follen, C.; Aragonés, J. Connectedness and well-being in simulated nature. *Appl. Psychol. Health Well Being* **2022**, *14*, 397–412. [CrossRef] [PubMed]
38. Martin, A.; Nejad, H.; Colmar, S.; Liem, G. Adaptability: How students' responses to uncertainty and novelty predict their academic and non-academic outcomes. *J. Educ. Psychol.* **2013**, *105*, 728–746. [CrossRef]
39. Strachan-Regan, K.; Baumann, O. The impact of room shape on affective states, heartrate, and creative output. *Heliyon* **2024**, *10*, e28340. [CrossRef]
40. Chin, T.-C.; Ricci, E.; Cooper, A.; Downie, A.; Vella-Brodrick, D. Using theory of change for fostering well-being and engagement in learning communities. In *Handbook of Positive Psychology in Schools: Supporting Process and Practice*, 3rd ed.; Allen, K.-A., Furlong, M.J., Vella-Brodrick, D., Suldo, S.M., Eds.; Routledge: Oxfordshire, UK, 2022; pp. 417–436. [CrossRef]
41. Heschong, L. Daylighting and human performance. *ASHRAE J.* **2002**, *44*, 65–67.
42. Haythornthwaite, C. Rethinking learning spaces: Networks, structures, and possibilities for learning in the twenty-first century. *Commun. Res. Pract.* **2015**, *1*, 292–306. [CrossRef]
43. Herrington, J.; Reeves, T.C.; Oliver, R. Authentic Learning Environments. In *Handbook of Research on Educational Communications and Technology*; Spector, J., Merrill, M., Elen, J., Bishop, M., Eds.; Springer: Berlin/Heidelberg, Germany, 2014. [CrossRef]
44. Hughes, H.; Franz, J.; Willis, J. (Eds.) *School Spaces for Student Wellbeing and Learning: Insights from Research and Practice*; Springer: Berlin/Heidelberg, Germany, 2019.

architecture

MDPI

Article

Promoting Mental Health Through Campus Landscape Design: Insights from New Zealand Universities

Yuqing He *, Jacky Bowring and Gillian Lawson

School of Landscape Architecture, Lincoln University, Christchurch P.O. Box 85084, New Zealand; jacky.bowring@lincoln.ac.nz (J.B.); gillian.lawson@lincoln.ac.nz (G.L.)
* Correspondence: yuqing.he93@gmail.com

Abstract: Mental health challenges among university students and staff are a pressing concern globally and in Aotearoa, New Zealand. Despite adopting frameworks like the Okanagan Charter to promote health and well-being, there is a lack of empirical research on how campus landscapes contribute to mental health promotion. This is a preliminary study based on a Ph.D. research project aiming to investigate the role of campus landscapes in supporting relaxation and internal recovery through everyday activities. We conducted a comparative multi-case study involving 66 participants from the University of Auckland, Lincoln University, and the University of Otago, exploring how they use and prefer campus landscapes for relaxation. Our findings indicate that 'enjoying nature' is the most preferred relaxation activity, with participants engaging both actively and passively with various spaces such as gardens, open lawns, and forested areas. Additionally, in campus settings, the proximity of relaxation spaces appears to be more important than design quality because of the limited time during working hours, which points to the importance of thoughtful campus planning. This study also found that university staff are often overlooked in discussions about healthy universities, despite their significant role in the campus setting. Overall, this study highlights the importance of biophilic design principles in creating health-promoting campus environments and offers initial insights for integrating natural elements into campus planning to enhance mental health and well-being.

Keywords: biophilic design; mental health; landscape architecture; campus; healthy university

1. Introduction

The intersection of mental health and university life has become increasingly prominent, with institutions often characterised as 'anxiety machines' [1–3]. Studies and reports identified concerns with students' mental health situation in multiple countries. For example, Lipson and colleagues analysed 10 years of data from the Healthy Minds Study and found significantly increased rates of mental health treatment and diagnosis among students in the United States [4,5]. The New Zealand Union of Students' Associations [6] produced the *Kei Te Pai?* report in 2018 (Kei Te Pai means fine or good in Te Reo Māori). It reviewed tertiary students' mental health and found that university students commonly experience moderate levels of psychological distress.

Similarly, in the past two decades, academic work has become more demanding. The main reason for this includes the dramatically increasing student numbers [7] and the students-as-consumer model derived from commercialised culture [8]. As a result, universities have then been characterised as "a generator of anxiety and pressure" [9]. Guthrie et al. [10], commissioned by the Royal Society and Wellcome Trust, explored the mental health of researchers, including academic staff and postgraduate students in universities, in the United Kingdom. They found that the level of job stress among university staff was comparable to that of healthcare workers, a high-risk group, with over 40 per cent of postgraduate students reporting symptoms of depression, emotional or stress-related problems, or high levels of stress. They also noted a lack of effective interventions and support for researchers, with even less literature evaluating the situation. Nicholls and colleagues conducted a meta-analysis of 26 papers and identified seven key themes of academic researchers' mental health experiences [11], highlighting that lack of job security coupled with high expectations has left researchers at risk of poor mental health and well-being.

However, universities need not be synonymous with stress; they can be places that foster health and well-being. This can be achieved not only through the curriculum but also by creating environments that promote healthy lifestyles. In 2015, the World Health Organization presented the Okanagan Charter, highlighting the role of higher education institutions in enhancing the health of those who live, learn, work, and play on campuses [12]. Universities are well-placed to educate students about healthy life choices and to instil the value of maintaining health and well-being in everyday life. They are also workplaces for staff who spend a significant portion of their daily lives there. As the Okanagan Charter has pointed out, work should be a source of health rather than consuming it.

The International Health Promoting Campuses Network has been established in 16 countries across five continents such as the African Health Promoting Campuses Network (AHPCN), the Tertiary Wellbeing Aotearoa New Zealand (TWANZ), the Asean University Network, the German Network of Health Promoting Universities (Arbeitskreis Gesundheitsfördernde Hochschulen), and the U.S. Health Promoting Campuses Network (USHPCN) [13]. Universities worldwide have gradually adopted the Okanagan Charter and prioritised health promotion in their agendas. However, as Travia et al. pointed out in their study in 2020, universities globally are still in the early stages of embracing well-being as one of their core objectives [14].

From a landscape architecture point of view, nature and landscapes have long been recognised to possess 'healing' power [15], and there has been considerable discussion about using natural landscapes to promote health and well-being in communities [16]. However, only limited research has examined the relationship between campus landscape design and health promotion. For example, Lau and Yang explored the application of healing gardens to campus design to create a health-supportive and sustainable campus environment in Hong Kong University [17]. Studies by Mt Akhir et al. mainly discussed the health effects of planting design on campus [18–20]. Holt et al. recognised the positive effects from social interactions during physical exercise sessions in green spaces, which they referred to as green exercises [21]. McDonald-Yale and Birchall explored winter design strategies that contribute to students' well-being for northern campuses [22].

Furthermore, studies on the health-promoting aspects of campus landscapes often lack a clear guiding concept or framework. One key design concept emerging from the field of restorative environments is biophilic design, which emphasises the integration of natural elements into built environments. Ulrich expanded on this concept, demonstrating that humans respond positively to certain natural elements—such as vegetation, water, and sunlight—which can have stress-relieving effects [23,24].

Over the past few decades, biophilic design has evolved into a structured framework comprising a series of values, principles, and attributes. Kellert identified seven key elements of biophilic design: environmental features, natural shapes and forms, natural patterns and processes, light and space, place-based relationships, evolved human–nature relationships, and over 70 associated design attributes [25]. While biophilic design provides broad design principles, applying these elements effectively in specific contexts, such as campus landscapes, remains a challenge. To date, little has been done to systematically integrate biophilic design into university settings for health-promoting purposes.

Overall, there remains a noticeable gap in empirical research on the role of campus landscapes in promoting mental health, particularly in New Zealand. Therefore, this study aims to fill this gap by investigating how campus landscape design supports relaxation for both students and staff. By exploring the uses and preferences of campus landscapes for relaxation, we seek to inform future designs of health-promoting campus environments.

2. Materials and Methods

This study is based on a Ph.D. research project and was given ethics approval by the Human Ethics Committee of Lincoln University, New Zealand (no. 2018-21). In-depth individual interviews were conducted during semester time in September and October 2018, with a total of 65 participants. This included 45 students, 16 academic staff, 2 administrative staff, and 2 medical staff from the city campus of the University of Auckland in Auckland, Lincoln University in Christchurch, and University of Otago in Dunedin (see Figure 1). Table 1 illustrates the key demographic characteristics of the participants.

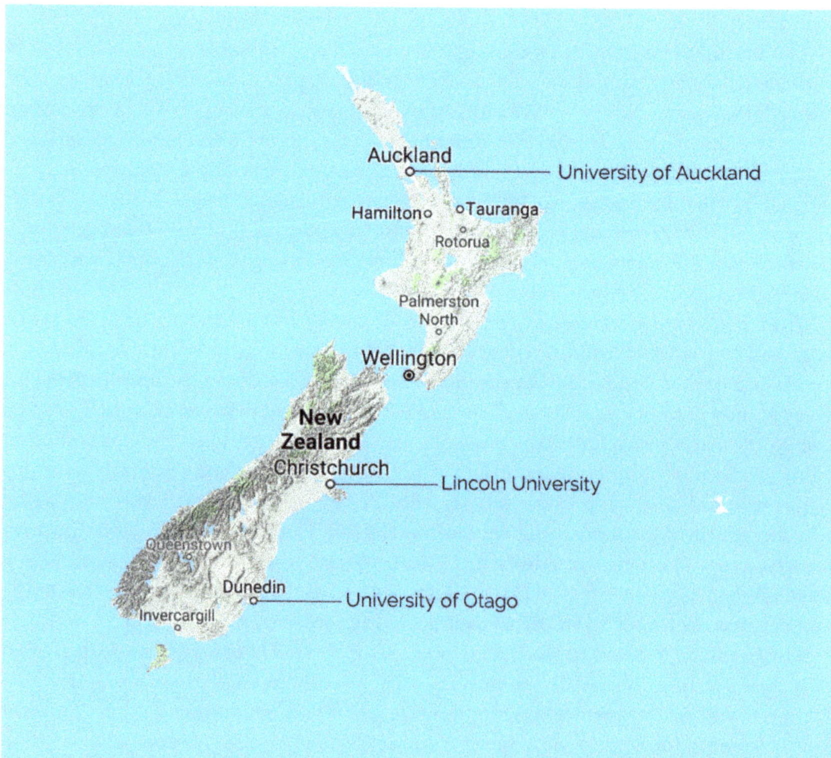

Figure 1. The main geographical location of the selected campuses (own work, 2020).

Table 1. Key demographic characteristics of participants (own work, 2020).

	The University of Auckland (*n*)	University of Otago (*n*)	Lincoln University (*n*)	%
Occupation				
- Student	20	16	11	72%
- Academic staff	6	6	2	22%
- Administrative staff	0	2	0	3%
- Health professionals	0	0	2	3%
Gender				
- Male	12	11	8	48%
- Female	14	13	7	52%
- Other	0	0	0	0
Age				
- Under 18	0	0	0	0
- 18–25	16	14	8	58%
- 25–50	6	6	6	28%
- Over 50	4	4	1	14%
Ethnicity				
- European	13	18	11	65%
- Maori	0	1	2	5%
- Asian	8	4	2	21%
- Pacific	1	1	0	3%
- Other	4	0	0	6%

This study drew on the methods used by previous studies; interviews and surveys are common methods used in restorative environment studies where they investigate participants' responses to videotapes or photos of landscapes [26–28]. However, the experience of an environment is multi-sensory, e.g., the temperature varies outside but it is stable in a room. It could be argued that traditional interviews that take place in an office may not be sufficient to collect explicit responses about the studied environment. Therefore, this study adopted a walking interview technique as an addition to traditional interviews to collect explicit responses about the studied environment from interviewees. Walking in the studied environments can provoke a sense of connection to the environment, which grants the researcher access to respondents' attitudes and knowledge about the environment and, thus, offers privileged insights [29]

Another interview technique adopted was the mental mapping exercise that asks respondents to draw and write about relaxation experience on campus. This technique is particularly helpful when respondents are not able to conduct a walking interview due to their availability. The action of drawing can encourage spatial thinking, which helps respondents to convey the everyday life image in their head.

Interviews that took place in the field offered the opportunity to conduct direct observation on site as well, which could help strengthen the data quality. Direct observation enabled evaluation of campus space use and justification of information provided by respondents from a landscape architecture professional perspective. Photographs of campus landscapes used for relaxation were taken; as Dabbs suggests, photographs of a site contain important characteristics of the studied site, which facilitates outside observers' understanding [30].

Data collection was carried out in the order of the University of Otago, Lincoln University, and the University of Auckland to make sure data were collected before their semesters ended when spaces were being used by campus users.

Each mental mapping exercise took around 5 to 15 min, which was a reasonable amount of time that campus users were willing to spend during their working day. Conversations with participants were recorded. Two open-ended questions were also presented to participants after they finished their drawing to encourage reflection and gain more insights into their relaxation experience and their perceptions of relaxation spaces on campus. The questions were as follows:

1. If you were going to use some keywords to describe the relaxation spaces you have drawn, what would they be?
2. What do you think are the most important factors for an outdoor space to be inviting/attractive for a university user to come and relax?

Each participant was asked if s/he was interested and available to conduct a further walking interview after finishing the mental mapping exercise. For those who were willing to participate, we arranged a time before the researcher left to avoid the difficulty of reaching participants again by email.

A total of 11 participants conducted the walking interview, with 6 of them being university staff. Others were not able to participate due to their availability. The walking interviews allowed participants to lead the way to the relaxation spaces they had identified on their maps. Detached from their workplaces, they were more willing to discuss campus relaxation spaces and even shared personal opinions on health and well-being. Being in the environment helped them recall their experiences on campus, and they often told stories of friends' or colleagues' relaxation experiences. Additionally, while on campus, participants could identify spaces they had overlooked and not included in their mental maps. For example, during one interview, as we stepped out of the building, a participant stopped and mentioned he had somehow omitted the courtyard—the closest space to his workplace—from his map. The walking interviews also provided opportunities for the researcher to share their direct observations about the campus and ask for their opinions.

The recorded conversations were transcribed alongside the information captured on the mental maps. Thematic analysis was employed to interpret the transcriptions, as this study aimed to develop a systematic understanding of relaxation on campus and offer insights for the design of future campus spaces. Nvivo 12 software was utilised to segment the conversations, summarise the content, and identify key themes and sub-themes. In this study, special attention was given to topics including the relaxation activities, the design characters of the identified relaxation spaces, and other relevant comments or evaluations of campus spaces for relaxation. Themes were reviewed and refined till they were informative and precise. Coding and producing themes were conducted for each case. The final analysis stage was the cross-case synthesis that compared the themes produced for each case to find commonalities and differences and examine plausible rival explanations. The finalised relaxation themes were then transferred to the map of each campus to show the design elements and landscape types that are associated with relaxation on campus.

3. Results

Campus users provided rich insights into their daily use and interpretation of campus landscapes for relaxation. This section highlights the main findings of this study:

1. Natural landscapes play a crucial role in campus relaxation, with 'nature' being the most mentioned keyword when describing relaxation spaces on campus.
2. In campus settings, practicality takes precedence. Campus users often have limited resources for relaxation during the day, making accessibility a priority. As a

result, proximity emerges as the most important factor in determining the use of a relaxation space.

3. University staff are challenging to engage, with a total of 18 participants being recruited and 6 of them participating in a walking interview. Despite the limited number of university staff, they offered detailed and thoughtful feedback. Due to the small number of staff interviewed, Section 3.3 is included to provide a more comprehensive understanding of campus environments for relaxation. Future studies may focus exclusively on this group, with greater effort made during the recruitment phase.

3.1. 'Nature'—The Top Keyword to Describe Relaxation Spaces on Campus

The thematic analysis revealed that natural elements were the most frequently mentioned keywords when the participants described their preferred relaxation spaces on campus. A student from the University of Auckland explained that "... because university life is always bustling and vibrant, that is why you come to parks or nature to get away and have a little breath". Over half of the participants included natural elements during their relaxation experience on campus through mental maps. According to the participants' accounts, the 'nature' they enjoy can be designed natural features such as trees, lawns, flowers or flower beds, and water features, or natural elements such as birds and even fresh air. For example, the mental map of a student participant from Lincoln University shows trees (see Figure 2), the mental map from another student from the same university shows lawn areas (see Figure 3), and the mental map from a student at the University of Otago points out that the grassy area in front of the Clock Tower is the best spot for watching flowers (see Figure 4).

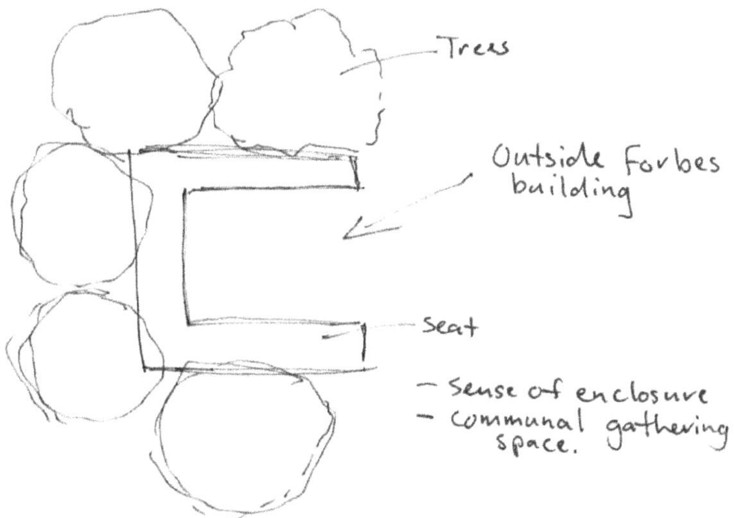

Figure 2. Participant LU15's mental map shows trees on Lincoln University campus (from participant LU15, 2018).

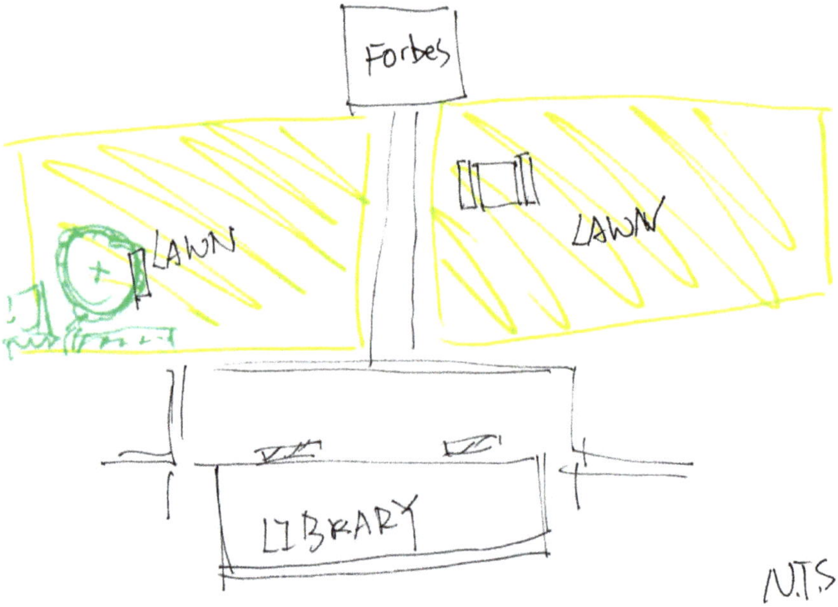

Figure 3. Participant LU6's mental map shows lawn areas at Lincoln University (from participant LU6, 2018).

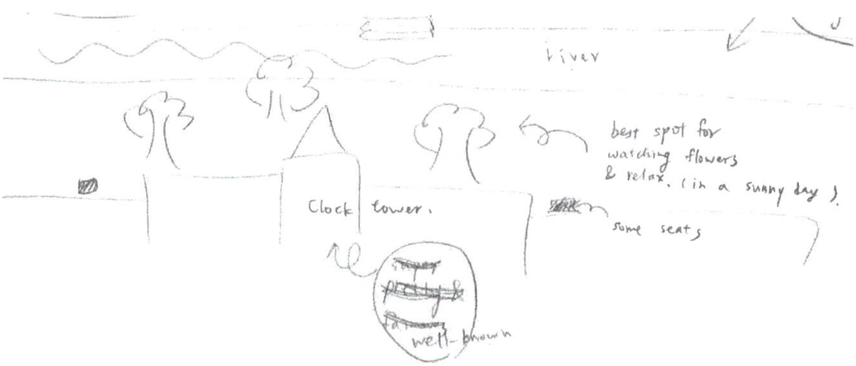

Figure 4. Participant UO21 identified a flower-watching spot on her mental map of the University of Otago (from participant UO21, 2018).

According to the participants, another benefit of nature is the sense of 'change' it provides, offering multi-sensory stimulation in contrast to the relatively static indoor environments where they spend much of their time. For example, they noted how the sun moves throughout the day, creating shifting shadows, whereas interior spaces are illuminated by fluorescent lighting, especially those with small windows. Many natural features also move with the wind and change with the seasons, such as leaves on deciduous trees, while interior spaces remain lifeless, filled with computers and work documents. A lecturer from Lincoln University explained that spaces with natural features make her feel like there are "lots of different things happening". She further explained:

> "I like looking at the change as well as what happens in the space. . . . I guess part of this relaxation for me is not necessarily always stopping. I feel like I can

recharge my batteries by moving through [spaces]. Like walking through here, that variation. Rather than at your office where you always got one window with nothing changes much throughout the day. . . . Sometimes you also notice when the flowers and the weather or the apples are coming in, so you kind of have that relationship [with the land]".

Participants from the University of Otago shared a variety of uses of the Leith River and the riverbanks for relaxation. Some participants simply enjoyed the environment brought by the fully landscaped riverbanks and flowering trees. According to the participants, they enjoy "getting some fresh air outside", "hearing birdsong and water rippling", and "seeing the changes of the season". Participants shared their own experience and observations of many campus users during spring. Residents who live in Dunedin, and even tourists, visit the Leith riverbanks in the warmer weather when the cherry and magnolia trees are in blossom (see Figure 5).

Figure 5. Campus users at the University of Otago enjoy natural features on the banks of the Leith River in more passive ways (photo by authors, 2018).

In addition to appreciating nature on campus, a few campus users also, though less frequently, found ways to actively engage with and utilise natural spaces such as playing sports. Based on the participants' accounts and our observations, many campus users enjoy using flat lawn areas as small-scale sports fields. For example, the researcher observed two people practising fencing during a weekend on the Leith riverbank at the University of Otago (see Figure 6), and some students set up a volleyball net and played at Lincoln University (see Figure 7).

Figure 6. Two people practised fencing on the banks of the Leith River at the University of Otago during a weekend (photo by authors, 2018).

Figure 7. Campus users play volleyball on the Forbes Lawn at Lincoln University (photo by authors, 2018).

An administrative staff from the University of Otago shared a story of the Leith River. "One time there was floodwater coming down at the Leith here, and people were kayaking in flood", she said with amusement and excitement, "I mean, they are not supposed to be kayaking and it is probably dangerous, but they really liked the water rapids!".

3.2. Proximity—The Most Important Factor for a Relaxation Space on Campus

While 'enjoying nature' is the most preferred relaxation activity on campus, both students and staff often find it challenging to do so due to the limited time for relaxation. A majority of participants agreed that proximity comes before the design quality of the space. For example, a student from the University of Auckland said: "I wouldn't want to go to OGGB [Sir Owen G Glenn Building] or Albert Park. But then I would use that outdoor area [the courtyard of the School of Architecture and Planning] even though it's not quite as nice just because I am here". His mental map further revealed the preference for proximity through his way of identifying relaxation spaces according to the adjacent building (see Figure 8).

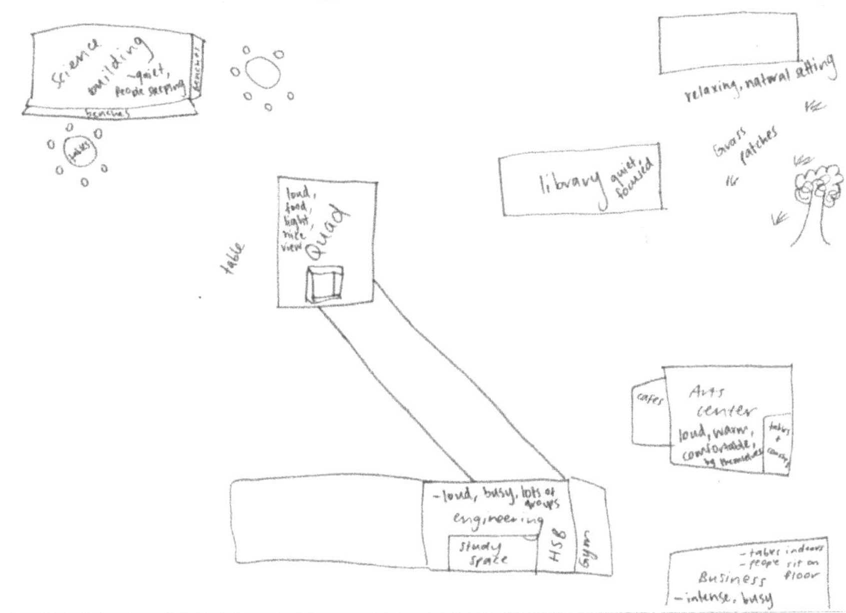

Figure 8. The importance of the context of a space at Lincoln University (from participant LU12, 2018).

Proximity is closely related to time and distance. The participants characterised their relaxation into two categories:

- Short-term breaks (less than 10 min);
- Long-term breaks (longer than 10 min).

3.2.1. Short-Term Breaks on Campus

There are three main types of spaces suitable for short-term breaks on campus: (1) communal space, (2) isolated space, and (3) walk-scape.

1. Communal Space

Outdoor spaces that are immediately accessible are utilised the most, even if they do not fully meet relaxation needs. Such proximity fosters and strengthens a sense of belonging, leading the participants to refer to these areas as 'communal spaces' since they are usually the downstairs spaces attached to buildings. These spaces are also sought after when campus users want to refresh their minds by simply peeking out of the window. For example, a student from the University of Auckland described the courtyard of the School of Architecture and Planning as follows:

> "It is in the intermediate zone. The people I would encounter there are the people who would belong to this large community, the Architecture and Planning, including all their friends or acquaintances. Generally, the main public would not walk through it because it is protected by buildings and the way it is shaping. But it is quite busy and communal".

She also showed the spatial character of the courtyard on the mental map (see Figure 9).

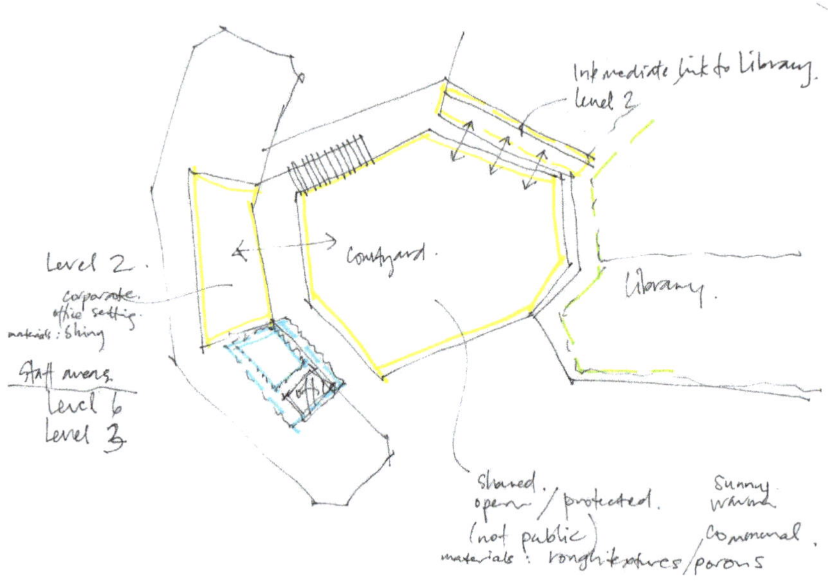

Figure 9. The spatial character of the courtyard of the School of Architecture and Planning, University of Auckland (from participant UA17, 2018).

2. Isolated Space

While communal spaces are often busy and noisy due to heavy use, people sometimes need a moment of quiet to recover and unwind. At such times, they might choose a more isolated space that is close to them and enjoy a semi-enclosed environment, such as the 'little forest' at Lincoln University. This 'little forest' is a small, lush, vegetated area located right beside the School of Landscape Architecture building (see Figure 10). Because of its proximity to the building, this isolated space can afford a quick refresh of the mind (see Figure 11).

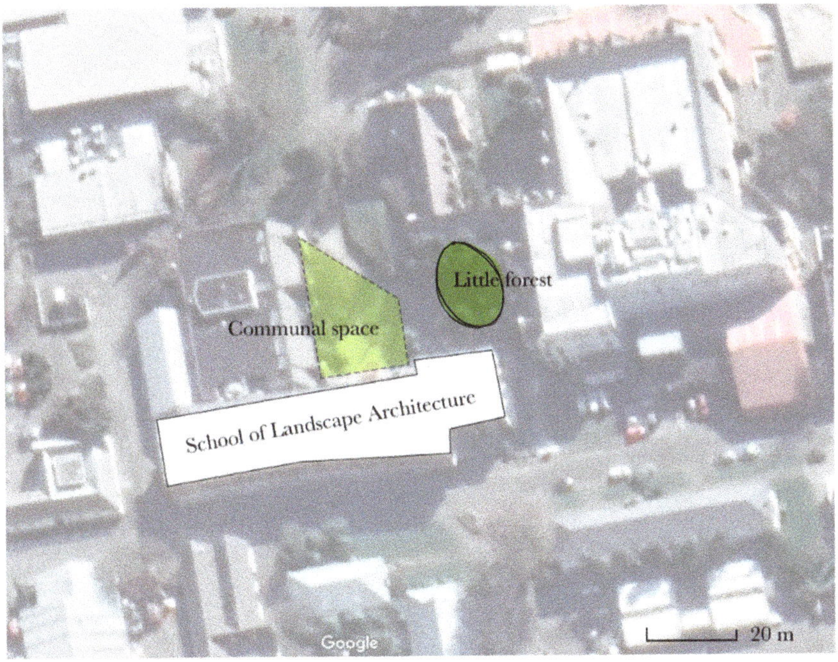

Figure 10. The location of the little forest at Lincoln University (modified from Google Map, 2020).

Figure 11. The view inside the little forest at Lincoln University (photo by authors, 2018).

3. Walk-scape

The last type of short-term relaxation space is not exactly a 'space' but rather the short journeys within campus, such as walking from one lecture theatre to another. The relaxation experience depends on the overall planning and design of campus streetscapes. A student from Lincoln University explained that he enjoys walking along campus streets several times a week to catch the bus, appreciating the blossoms that turn the avenue white and pink. A few participants from the University of Otago also expressed their appreciation of the innovative landscape work completed recently, noting that it promotes the consistency and aesthetics of pedestrian paths, creating a more walker-friendly environment (see Figure 12).

Figure 12. A fully paved campus street at the University of Otago (photo by authors, 2018).

3.2.2. Long-Term Breaks

The participants' accounts revealed that, on campus, the lunch break is almost synonymous with long-term breaks, and every campus in this study has multiple spaces specifically designed for lunchtime relaxation. Such spaces in all three campuses are open green spaces so that they can not only accommodate food service but also serve as centres for activities and student services (see Figure 13). Therefore, the participants characterised them as 'activity centres' and, because they are versatile and full of energy, they become 'go-to' spots where people not only satisfy their physical needs but also fulfil their social needs by meeting friends or occasionally joining fun activities. The participants suggested that during lunchtime, these spaces are likely the most bustling areas on the entire campus.

Figure 13. The main activity centres at (from top to bottom) (**a**) Lincoln University, (**b**) the University of Auckland, and (**c**) the University of Otago (photos by authors, 2020).

Aside from lunch breaks, campus users sometimes have time to relax between classes or after finishing their work for the day. A few participants prefer to use this time to go somewhere detached from the main campus atmosphere, which they called "get-away spaces". Participants explained that campus users need to temporarily escape from their work, social roles, and stress to simply be themselves and reconnect with nature and the larger world. For example, a health professional from Lincoln University referred to the magnolia garden as her "secret place" because "it is often just you there, and it is very peaceful". She further explained, "Even if other people are there, they are there for the same reason as you, so they are just looking for some quietness and some nature".

Some get-away spaces are located on campus and not far from activity centres. However, these spaces usually have stronger buffers than isolated spaces, creating a more enclosed environment. Get-away spaces on campus also offer no particular resources such as food or water, so campus users do not come to those spaces out of necessity. For example, the secret garden at Lincoln University is physically close to the major and secondary activity centres. However, with a long two-storey building blocking the view of the space from the main campus street, not many people gather there (see Figure 14).

Figure 14. The secret garden at Lincoln University blocked by the adjacent building block (photo by authors, 2019).

Some get-away spaces identified by the participants are located outside the main campus. By getting away from the main campus, the participants stated that they can forget about work, change mood instantly, and also feel more relaxed without social pressure. All get-away spaces identified by the participants that are outside the main campus are dominated by natural elements. For example, almost all participants from the University of Auckland enjoy having a large public park, Albert Park, located on the other side of the street from the main activity centre (see Figure 15), and a participant, a student from the University of Auckland, recommended Albert Park as "an escape from the campus".

Figure 15. Albert Park as one get-away space for campus users at the University of Auckland (photo by authors, 2018).

3.3. University Staff and Relaxation

The initial intention of this study was to investigate how both students and staff use campus landscapes for relaxation. Although the number of staff participants was limited, their perspectives were included to provide a more comprehensive understanding. As a preliminary study, this research serves as a foundation for future investigations, making it valuable to consider the insights and experiences of staff, even if the sample size was small.

Overall, university staff are more aware of the importance of relaxation during working hours. However, they appear to be at two extremes: they either have their own daily relaxation routines, or they are too busy to leave their desks.

One of the medical professionals from Lincoln University falls into the former category, and she likes

> ". . . doing qigong somewhere out of the way and a bit sheltered from the wind. This is not a place where you see many other people, but it is kind of a contemplative space where you can make strange movements like this because you do not have people walking by".

Another medical professional from Lincoln University likes the walking journey from the Recreation Centre on campus to the magnolia garden, which takes about 20 min one way.

In contrast, a staff member from the University of Auckland shared a markedly different experience, stating, "Lunch break is that little piece of desk between the laptop and the edge of the desk". While university staff may not always have the time to leave their office and relax, they can still benefit from brief moments spent looking outside the window—what they referred to as 'micro-breaks'. As a medical professional from Lincoln University pointed out:

> "... just look at the greenery around here, the flowers, and the change of the seasons. ... Sometimes we see people out there wine tasting or just having a discussion group. It's really nice. ... Sometimes I see people sitting on the bench, doing nothing, just relaxing, which I think is great. There's a really beautiful blossom tree and I just look up onto and see what the birds and the flowers and the leaves are doing, which is a relief for me". On the other hand, over half of university staff participants pointed out that university staff are often absent from some of the popular relaxation spaces identified by students, such as the activity centres, and there is no activity centres specifically designed for staff to relax. While campus spaces may appear to welcome all campus users, many participants—including both students and staff—pointed out that activity centres are more like "students' spaces".

Both groups explained that students experience a sense of belonging in places like activity centres, but there are no spaces specifically designed for university staff to foster the same feeling. Some academic staff expressed a desire to enjoy their lunch breaks somewhere away from students. As a lecturer from the University of Auckland explained:

> "Sometimes, if you want to go outside the office, you want to be detached from the emails and students. Sometimes I go to the Old Government Hall and have a glass of beer. When I'm so tired, such as on a Friday afternoon, I am looking for a place to escape".

4. Discussion

This study investigated how campus landscape design supports relaxation during working hours for both students and staff at New Zealand universities. Through mental mapping exercises and walking interviews with 66 participants across three universities, several key themes emerged: the significant role of natural landscapes in providing relaxation; the prioritisation of proximity over design quality due to time constraints; and the underrepresentation of university staff in the use of campus relaxation spaces.

4.1. Advocate for a Biophilic Campus

The study findings conclude that biophilic experiences dominate campus users' relaxation activities, with enjoying nature identified as their most preferred way to relax. This suggests that biophilic design can enhance the health and well-being potential of built environments, promoting people's health in everyday settings. Baur [31] and Liu et al. [32] further suggest that campus natural spaces can contribute to student success.

The enjoyment of nature on campus is primarily associated with the "direct experience of nature", one of the biophilic experience categories defined by Kellert and Calabrese [33]. According to the participants' accounts, this direct experience of nature is multi-sensory which attracts their attention away from work.

Participants often described the sight of nature as "beautiful", enhancing the aesthetics of outdoor landscapes and providing engaging window views. This finding aligns with numerous examples of research on the pleasure and stress reduction value of a window to natural landscapes (e.g., [34–37]). Although this study focused on landscape design strategies for outdoor spaces, the participants also expressed interest in the design of interior spaces for relaxation, which is pertinent given the significant amount of time they spend indoors. Although the micro-breaks mentioned in Section 3.3 fall outside the main scope of this study, it would be interesting for future research to explore the effects of interactions between interior and exterior spaces on relaxation.

Participants reported loving natural sounds such as birds chirping, wind rustling through leaves, and running water. Their preference for natural soundscapes over urban noises supports the idea that natural sounds can enhance people's experience of urban environments [38]. They also expressed enjoyment in engaging with nature when flowers are in bloom, relishing the pleasant fragrances.

The sight and smell of nature not only provide pleasure but also encourage tactile and even gustatory engagement, like picking fruit from a tree, fostering a sense of connection with nature and place [39]. Participants indicated that the multi-sensory experiences of nature are for more interesting than the static character of interior spaces. Kaplan and Kaplan termed this "soft fascination" which captures attention away from daily concerns in an "undramatic fashion" that "permits a more reflective mode" [40].

This multi-sensory relaxation experience aligns with Thompson's findings, where woodlands were places for children and parents to experience extraordinary, sensory, and emotional encounters [41].

Moreover, biophilic design acknowledges the intrinsic relationship between humans and nature. It not only proposes integrating natural elements into built environments but also aims to identify design features and spatial configurations that align with our innate biophilic preferences, including those we try to avoid [15]. To date, little has been done to incorporate biophilic design into university settings for health-promoting purposes. Peters and D'Penna discussed trends and gaps in understanding the influence of biophilic design in university settings, highlighting the complexity of these environments and urging that biophilic elements be tailored to meet users' needs [42]. For instance, while the participants in our study identified the 'little forest' as a relaxation space, considerations of safety should be addressed in future development.

4.2. *Advocate for Relaxation-Oriented Campus Planning*

The participants' accounts revealed that proximity is the priority when selecting a space for relaxation, even if the design features do not fully meet their needs. This indicates there is a fundamental difference between relaxation during working hours and relaxation after working hours. Kaplan and Kaplan also recognised proximity as essential for nearby nature, providing convenient relaxation at people's doorsteps and facilitating a quick switch between work and rest; people tend to use what is readily available [40].

Campus users demonstrated an even stricter perception of proximity compared to users of community open spaces. They favoured communal spaces attached to buildings—like the front or back gardens of private properties—and spaces connected to facilities. In contrast, Sugiyama et al.'s study showed adult participants' walking activities are within a 1.6 km radius [43], which is significantly different from campus users' preferences.

This preference for immediately available relaxation spaces highlights the importance of incorporating relaxation into campus planning, enabling users to take short breaks and prevent burnout. Moreover, many studies of public open spaces focus on the walking experience, suggesting that if people intend to walk, distance is less of a barrier. Therefore, it would be interesting to explore whether distance becomes more critical for landscape users engaging in other types of relaxation activities in community spaces during non-working hours.

4.3. *Advocate for Healthy Workplaces*

Universities, as large employers, have the capacity to provide a caring and supportive working environment for staff [44]. However, the participants' accounts from this study revealed that campus planning and landscape design often lack the attention given to university staff. With academic work becoming increasingly demanding, campus landscapes should include spaces specifically designed for staff to relax during working hours and prevent burnout.

Geurts and Sonnentag defined recovery during working hours as internal recovery, and recovery after hours as external recovery—including after work, during weekends, or on holidays [45]. They argued that the need for external recovery increases when internal recovery is insufficient, suggesting that internal recovery can buffer accumulated fatigue and stress. Therefore, it is essential to explore how workspaces can be designed to promote internal recovery and help maintain the 'healthy' mental status of workers.

Some studies have shown that internal recovery associated with landscapes close to workplaces can deliver mental health benefits. For example, the landscape architect Stigsdotter explored employees' experiences of stress and their use of green outdoor spaces at their workplaces in nine Swedish cities. She found that accessible green spaces adjacent to workplaces can positively influence employees, even if they only look at the green view [46].

5. Conclusions

Through investigating the uses of campus landscapes for relaxation and the preferred relaxation spaces at the University of Auckland, Lincoln University, and the University of Otago, relaxation associated with natural landscapes turned out to be the most preferred form of relaxation on campus. Natural landscapes on campus are not untouched 'wild' nature but manmade natural features. They provide spaces for 'enjoying nature' during short-term and long-term breaks. The enjoyment of nature on campus for relaxation recognises the health and well-being potential of biophilic design and suggests that integrating natural features in the built environments can enhance people's health and well-being in everyday settings. In addition, the time available for relaxation is the major difference between internal and external recovery. Therefore, proximity comes before the design quality of space when choosing spaces for internal relaxation. These findings not only contribute to designing healthy campus environments for living, learning, and working, but also to workspace design in cities.

Author Contributions: Writing—original draft, Y.H.; Writing—review & editing, J.B. and G.L. All authors have read and agreed to the published version of the manuscript.

Funding: This research received no external funding.

Data Availability Statement: The original contributions presented in this study are included in the article. Further inquiries can be directed to the corresponding author(s).

Conflicts of Interest: The authors declare no conflict of interest.

References

1. Hall, R.; Bowles, K. Re-engineering Higher Education: The Subsumption of Academic Labour and the Exploitation of Anxiety. *Workplace A J. Acad. Labor* **2016**, *28*, 30–47.
2. Institute for Public Policy Research. Available online: https://www.ippr.org/articles/not-by-degrees (accessed on 27 January 2025).
3. Prince, J.P. University student counselling and mental health in the United States: Trends and challenges. *Ment. Health Prev.* **2015**, *3*, 5–10. [CrossRef]
4. Healthy Minds Network. Available online: https://healthymindsnetwork.org/wp-content/uploads/2021/02/HMS-Fall-2020-National-Data-Report.pdf (accessed on 25 November 2024).
5. Lipson, S.; Lattie, E.; Eisenberg, D. Increased rates of mental health service utilization by U.S. college students: 10-year population-level trends (2007–2017). *Psychiatr. Serv.* **2018**, *70*, 60–63. [CrossRef] [PubMed]
6. New Zealand Union of Students' Associations. Available online: https://kimspaincounselling.files.wordpress.com/2019/05/student-mental-health-report-kei-te-pai.pdf (accessed on 29 November 2024).
7. Kinman, G. Doing More with Less? Work and Wellbeing in Academics. *Somatechnics* **2014**, *4*, 219–235. [CrossRef]
8. HEPI. Available online: https://www.hepi.ac.uk/wp-content/uploads/2019/05/HEPI-Pressure-Vessels-Occasional-Paper-20.pdf (accessed on 29 November 2024).

9. Education Support. Available online: https://www.educationsupport.org.uk/media/fs0pzdo2/staff_wellbeing_he_research.pdf (accessed on 29 November 2024).
10. Guthrie, S.; Lichten, C.A.; Van Belle, J.; Ball, S.; Knack, A.; Hofman, J. Understanding mental health in the research environment: A Rapid Evidence Assessment. *Rand Health Q.* **2018**, *7*, 2.
11. Nicholls, H.; Nicholls, M.; Tekin, S.; Lamb, D.; Billings, J. The impact of working in academia on researchers' mental health and well-being: A systematic review and qualitative meta-synthesis. *PLoS ONE* **2022**, *17*, e0268890. [CrossRef]
12. World Health Organization. Available online: https://www.who.int/teams/health-promotion/enhanced-wellbeing/first-global-conference (accessed on 29 November 2024).
13. International Health Promoting Campuses Network. Available online: https://www.healthpromotingcampuses.org/networks (accessed on 22 January 2025).
14. Travia, R.M.; Larcus, J.G.; Andes, S.; Gomes, P.G. Framing well-being in a college campus setting. *J. Am. Coll. Health* **2020**, *70*, 758–772. [CrossRef]
15. Ulrich, R.S.; Gilpin, L. Healing arts: Nurtrition for the soul. In *Putting Patients First: Designing and Practicing Patient-Centered Care*; Frampton, S.B., Charmel, P., Charmel, P.A., Eds.; Jossey-Bass: San Francisco, CA, USA, 2003; pp. 117–146.
16. Maas, J.; Verheij, R.A.; Groenewegen, P.P.; de Vries, S.; Spreeuwenberg, P. Green space, urbanity, and health: How strong is the relation? *J. Epidemiol. Community Health* **2006**, *60*, 587–592. [CrossRef]
17. Lau, S.S.Y.; Yang, F. Introducing Healing Gardens into a Compact University Campus: Design Natural Space to Create Healthy and Sustainable Campuses. *Landsc. Res.* **2009**, *34*, 55–81. [CrossRef]
18. Mt Akhir, N.; Md Sakip, S.R.; Abbas, M.Y.; Othman, N. Landscape Spatial Character: Students' preferences on outdoor campus spaces. *Asian J. Qual. Life* **2018**, *3*, 89–97. [CrossRef]
19. Mt Akhir, N.; Md Sakip, S.R.; Abbas, M.Y.; Othman, N. A Taste of Spatial Character: Quality outdoor space in campus landscape leisure setting. *Environ.-Behav. Proc. J.* **2017**, *2*, 65–70. [CrossRef]
20. Mt Akhir, N.; Sakip, S.R.M.; Abbas, M.Y.; Othman, N. Literature survey on how planting composition influence visual preferences: A campus landscape setting. *J. Soc. Sci. Res.* **2018**, *6*, 783–790. [CrossRef]
21. Holt, E.W.; Lombard, Q.K.; Best, N.; Smiley-Smith, S.; Quinn, J.E. Active and Passive Use of Green Space, Health, and Well-Being amongst University Students. *Int. J. Environ. Res. Public Health* **2019**, *16*, 424. [CrossRef] [PubMed]
22. McDonald-Yale, E.; Birchall, S.J. The built environment in a winter climate: Improving university campus design for student wellbeing. *Landsc. Res.* **2021**, *46*, 638–652. [CrossRef]
23. Ulrich, R.S. Biophilia, Biophobia, and Natural Landscapes. In *The Biophilia Hypothesis*; Kellert, S.R., Wilson, E.O., Eds.; Island Press: Alameda, CA, USA, 1993; pp. 73–137. [CrossRef]
24. Ulrich, R.S. Biophilic theory and research for healing design. In *Biophilic Design: The Theory, Science, and Practice of Bringing Buildings to Life*; John Wiley & Sons, Inc.: Hoboken, NJ, USA, 2008; pp. 87–106.
25. Kellert, S.R. Dimensions, Elements, and Attributes of Biophilic Design. In *Biophilic Design: The Theory, Science, and Practice*; Kellert, S.R., Heerwagen, J.H., Mador, M.L., Eds.; John Wiley & Sons, Inc.: Hoboken, NJ, USA, 2008; pp. 3–19.
26. Plane, J.; Klodawsky, F. Neighbourhood amenities and health: Examining the significance of a local park. *Soc. Sci. Med.* **2013**, *99*, 1–8. [CrossRef]
27. Purani, K.; Kumar, D.S. Exploring restorative potential of biophilic servicescapes. *J. Serv. Mark.* **2018**, *32*, 414–429. [CrossRef]
28. Ulrich, R.S.; Simons, R.F.; Losito, B.D.; Fiorito, E.; Miles, M.A.; Zelson, M. Stress recovery during exposure to natural and urban environments. *J. Environ. Psychol.* **1991**, *11*, 201–230. [CrossRef]
29. Ingold, T.; Lee, J. Introduction. In *Ways of Walking: Ethnography and Practice on Foot*; Vergunst, J.L., Ingold, T., Eds.; Routledge: London, UK, 2008; pp. 1–20.
30. Dabbs, J.M. Making Things Visible. In *Varieties of Qualitative Research*; Van Maanen, J., Dabbs, J.M., Faulkner, R.R., Eds.; SAGE Publications Ltd.: Beverly Hills, CA, USA, 1982; pp. 31–63.
31. Baur, J. Campus community gardens and student health: A case study of a campus garden and student well-being. *J. Am. Coll. Health* **2020**, *70*, 377–384. [CrossRef]
32. Liu, W.; Sun, N.; Guo, J.; Zheng, Z. Campus Green Spaces, Academic Achievement and Mental Health of College Students. *Int. J. Environ. Res. Public Health* **2022**, *19*, 8618. [CrossRef]
33. The Practice of Biophilic Design. Available online: https://www.biophilic-design.com/ (accessed on 29 November 2024).
34. Bardwell, L.V. *Nature Around the Corner: Preference and Use of Nearby Natural Areas in the Urban Setting*; University of Michigan: Ann Arbor, MI, USA, 1985.
35. Kardan, O.; Gozdyra, P.; Misic, B.; Moola, F.; Palmer, L.J.; Paus, T.; Berman, M.G. Neighborhood greenspace and health in a large urban center. *Sci. Rep.* **2015**, *5*, 11610. [CrossRef]
36. Kearney, A.R.; Winterbottom, D. Nearby nature and long-term care facility residents: Benefits and design recommendations. *J. Hous. Elder.* **2006**, *19*, 7–28. [CrossRef]
37. Ulrich, R.S. View through a window may influence recovery from surgery. *Science* **1984**, *224*, 420–421. [CrossRef] [PubMed]

38. Hedblom, M.; Heyman, E.; Antonsson, H.; Gunnarsson, B. Bird song diversity influences young people's appreciation of urban landscapes. *Urban For. Urban Green.* **2014**, *13*, 469–474. [CrossRef]
39. Beatley, T. *Handbook of Biophilic City Planning and Design*; Island Press: Washington, DC, USA, 2016.
40. Kaplan, R.; Kaplan, S. *The Experience of Nature: A Psychological Perspective*; Cambridge University Press: Cambridge, UK, 1989.
41. Thompson, C.W.; Aspinall, P.; Bell, S.; Findlay, C. "It gets you away from everyday life": Local woodlands and community use—What makes a difference? *Landsc. Res.* **2005**, *30*, 109–146. [CrossRef]
42. Peters, T.; D'Penna, K. Biophilic design for restorative university learning environments: A critical review of literature and design recommendations. *Sustainability* **2020**, *12*, 7064. [CrossRef]
43. Sugiyama, T.; Francis, J.; Middleton, N.J.; Owen, N.; Giles-Corti, B. Associations between Recreational Walking and Attractiveness, Size, and Proximity of Neighborhood Open Spaces. *Am. J. Public Health* **2010**, *100*, 1752–1757. [CrossRef]
44. Tsouros, A.D.; Dowding, G.; Thompson, J.; Dooris, M. *Health Promoting Universities: Concept, Experience and Framework for Action*; World Health Organisation Regional Office for Europe: Copenhagen, Denmark, 1998.
45. Geurts, S.A.E.; Sonnentag, S. Recovery as an Explanatory Mechanism in the Relation between Acute Stress Reactions and Chronic Health Impairment. *Scand. J. Work. Environ.* **2006**, *32*, 482–492. [CrossRef]
46. Stigsdotter, U.K. A Garden at Your Workplace May Reduce Stress. In Proceedings of the Design & Health World Congress & Exhibition, Montreal, Canada, 25–29 June 2003.

architecture

MDPI

Article

Teacher and Parent Perception of Biophilic Conditions in Primary-School Environments and Their Impact on Children's Wellbeing

Bethania Lanzaro * and Marcella Ucci *

UCL Institute for Environmental Design and Engineering, The Bartlett Faculty of the Built Environment, University College London (UCL), London WC1H 0NN, UK
* Correspondence: bethania.lanzaro.18@alumni.ucl.ac.uk (B.L.); m.ucci@ucl.ac.uk (M.U.)

Abstract: The term "biophilia" refers to the intrinsic affinity that humans have towards nature, natural elements and natural processes. Biophilic design theories suggest that the introduction or representation of natural characteristics or elements into the built environment can help enhance people's health and wellbeing. Primary school buildings are important environments where children spend considerable time. However, there is limited evidence on the impact of their biophilic features on the children themselves and on perceptions of important facilitators of children's wellbeing, such as teachers and parents. This research aims to investigate whether teachers and parents perceive children to have a preference or desire for specific biophilic characteristics in their school's physical environment; and whether teachers perceive some biophilic characteristics as having an effect on children's performance and behaviour. A framework for evaluating biophilic characteristics in primary schools was developed. Two case study primary schools in London and Bath (England, UK) were audited against this framework, and teachers and parents were surveyed. The results suggest that children do have a preference towards the specific biophilic features studied, which is stronger and more demanding when the exposure is higher. For some aspects, teachers' perception of benefits is also susceptible to the quality of the environment itself.

Keywords: biophilia; primary school environment; children's wellbeing; perception; performance; behaviour

1. Introduction

Evolutionary theories suggest that humans are physiologically and psychologically adapted to natural environments since they emerged and long evolved within nature [1], creating adaptive responses to certain stimuli and configurations, mainly regarding food, water and security [2,3].

In their search for comfort and security, humans learned to modify their own physical environment. Rapid industrialisation and the massive migration to cities have precipitated these transformations, leading to highly modified environments which are more comfortable than our original habitat, but also resulting in a disconnection from nature to which we were originally adapted, physiologically and psychologically.

In environmental sciences, "biophilia" explains the intrinsic affinity that humans have towards nature, natural elements and natural processes. It has been proven that introduction or representation of natural characteristics or elements into the built environment can help enhance people's health and wellbeing within our contemporary habitat (e.g., the use of daylight, natural materials or images of nature).

Children, in particular, could be more sensitive than adults to their environment. According to the World Health Organization [4], children consume more air, water and food than adults on a weight proportion; several of their body systems are still growing and therefore very susceptible; they have different behavioural patterns towards the environment which can increase their exposure (to particles and chemicals in dust and soil); and children do not usually control their own environment, since adults do.

Schools are the environment where contemporary children spend up to 60% of their time. Poor school environments can impact children's and teachers' attendance and retention, and therefore also academic performance [5]. The concept of 'school environment' encompasses "physical environment", "educational environment" and "social environment" [6]; however, this research focuses mainly on physical conditions.

From an environmental psychology point of view, the design of the physical environment is believed to affect children's perception, learning and behaviour; and conditions the evolution of motor, cognitive and social skills [7]. An extensive post-occupancy evaluation on UK primary schools by the RIBA [5] found that "more than 9 in 10 teachers believe school design is important, 1 in 5 teachers have considered quitting because of the condition of school buildings and 91% of teachers feel good design is important to good pupil behaviour". More recently, Barrett et al. [8] evaluated the holistic impact of several aspects of primary-school buildings on children's learning performance, highlighting the importance of optimal physical environments.

1.1. Biophilia

Humans have an affinity to those elements and environments which have allowed our development and survival, including sources of water and some landscapes, such as savannah-type settings, forest edges, watercourses, mountains and valleys [2,9]. The Prospect-Refuge Theory proposed by Appleton [10] emphasises the psychological affinity humans have for wideness (prospect) and hiding possibilities (refuge), which emerges from our ancestors' safe exploration and wayfinding. As diurnal species, vision developed as our prominent sense, and is the reason why daylight, views, the possibility to see far, and certain colours and patterns are especially meaningful at an unconscious level [9].

The existing literature has proved the positive effect of the exposure to plants and other elements of nature on emotion [11], physiology [12–15], cognition [16–19], behaviour [20–22] and health [23–26].

There is worldwide research on children's environments related to biophilia. Rice and Torquati [27] surveyed 114 pre-scholars on "preferences for play locations, enjoyment of sensory aspects of nature, exploring nature, and curiosity about nature". Maller [28] studied the importance of children's "hands-on contact to nature" interviewing educators, school principals, and environmental-education-industry professionals, who recognised benefits on "self-esteem, engagement with school and a sense of empowerment". Similarly, Lee and Park [1] evaluated which biophilic design characteristics could apply to a children's library through case studies and surveys, concluding that the needs of these environments are to provide the experience of nature, shelter and open spaces, multi-functional spaces, and to promote sensorial experiences.

1.2. Nature and Restorativeness

The Attention Restoration Theory (ART), introduced by Kaplan and Kaplan in 1995 [29], proposes the concept of 'restorative environment' as that which "promotes (and not merely permits) restoration" [30] (p. 273). Cognitive functioning was proved to be restored with simple interactions with nature, comparing the cognitive response of young adults after a 50–55-min walk through an arboretum to a traffic-heavy street, and after 10 min of watching an image of a natural landscape to an urban landscape [31]. Children exposed to the natural environment presented better and faster responses [32], giving evidence of an association between cognitive restoration and eye focus direction.

In medicine, early studies associated views to nature/quality of the view with patients' faster recovery in hospitals [24,33]. More recent research suggests links between views to nature and the reduction of pain [34,35]. Other studies associate this link to stress recovery [36].

Bagot, Allen and Toukhsati summarise evidence proposing four requirements for an environment to be restorative: "being away (physical or psychological)", "compatibility (with purpose)", "fascination (cognitive or physical)", and "extent (scope and connectedness)" [37]. They examined and scored four studies and concluded that the indoor environment was less restorative than outdoor/natural, with the amount of vegetation being the only meaningful variable determining restorativeness, and that adults' and children's restoration might not follow the same environmental determinants.

1.3. Nature Deficit in Contemporary Children

Contemporary children might be suffering from what Louv calls "nature-deficit disorder": an insufficiency of 'primary experiences' in nature, crucial for their development. Multisensory experiences in nature, quietness and 'space to wander' are his highlighted aspects. He also suggests that one of the main causes of nature-deficit is parents' fear of safety [38].

Urban children tend to be more cooperative and creative in their play when exposed to natural settings, in comparison to spaces with less trees and grass or manufactured play areas [21,38]. Chawla et al. found that children identify natural spaces with freedom, happiness and opportunities to confront challenges [39].

Within the learning environment, school playgrounds with the presence of greenery work as refuge for stress-reduction, restoration and enhance children's social skills. Exposure to nature is proved to be related also to benefits such as "improved focus, vitality, productivity and reduced stress, factors that may enhance the academic performance of children" [40] (p. 16). Views of nature in classrooms were found to generate a positive perception of courses and better grades [41,42]. Han [43] argues that students are vulnerable to suffering from mental fatigue, this being the reason why their "ideal learning environment should promote attention focusing, reduce mental fatigue and psychophysiological stress, and ideally even improve health and encourage better learning" [43] (pp. 659–660). In his study, Han measured the subjective and objective influence and variations of high-school students during one semester in a classroom with plants. Using a control group, the students were surveyed, and their examination scores and behavioural records were collected. They concluded that "six limitedly visible plants in the classroom had an immediately significant and positive influence on the students' perceptions of preference, comfort and friendliness" [37] (p. 680). It is relevant to consider that in this type of study there is a possible Hawthorne Effect, where the impact of a positive response may occur only due to the modification of the environment, independently from the characteristics of the modification itself [44] which could play a role as well.

1.4. Existing Frameworks and Classifications for Assessing Biophilia in the Built Environment

In the last two decades, several authors have presented classifications which describe and synthesise those characteristics that define, in their view, biophilic design. These are interpretations of the built environment, where similar strategies and attributes are presented, with some variations in the categories used.

Kellert, also known as the 'godfather of biophilic design', presented several frameworks throughout his career. The first one, from 2002 [45], defined three categories for experiencing nature: "direct experience", which is related to non-human natural settings; "indirect experience", to refer to human-made natural environments; and "symbolic experience" for the appreciation of nature through books and television. The first two classifications were later used by himself with a different meaning, whereas the appreciation of nature through books and television was not mentioned again by other frameworks.

In a totally different approach, in 2008 [46] Kellert identified 70 "biophilic design attributes", which were classified into six strategies: "environmental features"; "natural shapes and forms"; "natural patterns and processes", with attributes such as "growth and efflorescence" and "dynamic balance and tension"; "light and space", with seven different attributes related to light and five to space; "place-based relationships", with attributes such as "avoiding placelessness" or "spirit of space"; and "evolved human-nature relationships". With this framework, Kellert proposes that the representation of nature and sense of belonging to the place can play a role for biophilia as well, which goes beyond the presence of natural elements per se.

In the same publication, Heerwagen and Gregory [47] examined seven aesthetic qualities from nature which could be evoked by architecture. New proposed attributes are "motion" and "sense of freeness". The authors claimed that nature is always on the move and that people enjoy watching movement, including watching people move in an urban landscape. Regarding "sense and freeness", the authors mention two particular aspects: the psychological and physical sense of freedom generated by the lack of boundaries between spaces, and the simple action of being able to open a window in one's environment, signifying control and therefore freeness. Some attributes from this framework have similarities with Kellert's from the same publication.

Terrapin Bright Green developed in 2014 [48] an extensive research based on the work of Kellert, Alexander, Heerwagen, Kaplan, Kaplan and Ulrich, among others. Here they presented 14 "patterns of biophilic design" under the categories "nature in the space", "natural analogues" and "nature of the space". One of the main novelties of this framework is that for the first time, the attributes (or patterns) are mapped against their effect on people's health and wellbeing. They classified the evidence into three types: stress reduction, cognitive performance and emotion, mood and preference.

In his last book [9], Kellert presented and described a compacted list of 25 "attributes of biophilic design". This framework brings the names of the categories from his first classification of 2002 but framed differently. For example, "direct experience" refers to elements from nature and not necessarily "non-human natural spaces", and "indirect experience" does not refer to botanical gardens of cultivated crops, but to representations of nature such as images, materials, and the simulation of natural elements.

In the same year, biophilic design entered into the field of building certifications. The International Living Future Institute developed a Biophilic Design Toolkit [49], as part of the requirements for the Living Building Challenge Certification. The categories and attributes used reframe Kellert's, proposing a series of design guidance and tools to implement them at the different project design phases. Simultaneously, the WELL Building Standard [50], which evaluates the impact of buildings on the health and wellbeing of its occupants, also included a few aspects of biophilia under the categories Mind and Light, some as "preconditions" and others as "optimisations". This is the first framework that introduces criteria for sufficiency.

Overall, the existing frameworks share similar strategies and attributes, with a clear tendency to use the concepts of "direct experience", "representation" and "spatial configurations". Except for WELL, which only considers very few aspects of biophilia, existing frameworks work as classifications in the way of checklists without providing priorities or hierarchies. They do not argue whether, for example, the direct presence of daylight or plants have a stronger impact than daylight simulation or nature-like patterns. Most frameworks also have no guidance to sufficiency, with little evidence on how biophilia can be achieved. And there is limited discussion on differences in populations. For instance, in existing frameworks, no particular consideration is taken for the way children perceive or experience the built environment, the physical spaces used mainly by children or how the effect of the built environment on children might vary in comparison to adults. Furthermore, there is limited research on perceptions of parents and teachers on the impact and role of biophilic features in school environments on children. These views are important when considering that parents' and teachers' actions and beliefs can directly or indirectly affect children's wellbeing, for example via impact on exposures (e.g., access to daylight and views might be controlled by teachers in classrooms) or child behaviours or interests.

1.5. Aims and Objectives

The aim of this study is to investigate whether teachers and parents perceive children as having preference or desire for specific biophilic features in their school's physical environment, and whether teachers perceive certain biophilic characteristics as affecting children's performance and behaviour.

This study focuses on those biophilic aspects easily identifiable by non-specialist publics. Likewise, 'preference', 'performance' and 'behaviour' are commonly used and widely understood terms that were selected as easily recognisable aspects related to well-being. Teachers' opinions are particularly relevant since they have some control over the classroom's environment (e.g., windows and blinds); and parents' opinions are relevant as they can reflect on what children discuss about their school environment at home.

Specifically, the objectives were as follows:

1. To develop a framework to evaluate biophilic aspects of school environments which allows categorising, comparing, and contrasting different school environments in a way which could be easily communicated with parents and teachers.
2. To evaluate teachers' perceptions about the role of specific biophilic features in children's performance and behaviour at school.
3. To identify aspects of children's preferences towards biophilic features, as perceived by parents and teachers.

2. Methods

2.1. Study Design Overview

The methodology followed involved:

1. The formulation of a bespoke framework for evaluating biophilic conditions in urban primary schools.
2. Audits of two case studies against this framework.
3. Surveys of teachers and parents about their perception about children's preference, performance and behaviour in relation to certain biophilic attributes. Due to ethics and data protection, participants were not asked to disclose details about specific children or health conditions.

2.2. Development of a Framework for Evaluating Biophilic Conditions in Urban Primary Schools

Based on the existing attributes and classifications from existing frameworks presented in Section 1.4, a new framework with a focus on children and primary school environments was formulated.

A concise selection of practical attributes, which could be suitable for assessing the primary-school physical environment, were identified and classified under "elements" or "spatial experiences", and "direct experience" or "representation" (see Table 1).

Table 1. Summary of attributes selected and the proposed classification.

	Presence of Elements	Spatial Interactions
Direct experience	Natural light, fresh air, water, plants, animals, natural materials.	Views out, infrastructure for outdoor classes, promotion of healthy commuting, gardening sessions, 'forest school' activities.
Representation	Images of nature, textures and patterns, colour palette, shapes and forms.	Spatial variability, prospect, refuge, active design, possibility to run, possibility to climb.

For each of these attributes, criteria were defined to determine what was considered as "achieved", "partially achieved", or "not achieved" (see Table 2). This framework does not aim to define minimum performance grades for biophilia (further research is needed to tackle this); nevertheless, it gives a basic evaluation of the school's biophilic condition.

Table 2. Criteria to evaluate levels of fulfilment for each attribute.

Attribute	Achieved	Partially Achieved	Not Achieved
Natural light	Daylight always present when weather allows it. Most times there are no problems with glare.	Not enough daylight/ daylight is not always present. Too much glare which results in the blinds being down for long periods.	Almost no daylight in the space.
Fresh air	Both conditions are met: 1. Natural ventilation. 2. Air perceived as fresh, not being stuffy, and not having disgusting smells.	Only one of the previous conditions are met.	None of the previous conditions are met.
Water	Any presence of water features (e.g., fountains, constructed wetlands, ponds, swales, etc.) available for all children.	Presence of water features only for some children. A water tap in the playground to which children have access can count.	No presence of any source of water (except in toilets).
Plants	Both conditions are met: 1. Indoor plants are present in all classrooms. 2. Outdoor plants are a dominant element in the playground.	Any of the following conditions are met: 1. Existing indoor plants but not in all classrooms. 2. Outdoor plants exist but are not a dominant element in playground.	Both conditions are met: 1. No indoor plants. 2. Outdoor plants are scarce (e.g., isolated planters) or non-existent.
Animals	Both conditions are met: 1. All children have access to pet animals (e.g., fish bowl). 2. It is easy to encounter insects or other wild animals (e.g., if there are planters).	Any of the following conditions are met: 1. Only few children have access to pet animals (e.g., fish bowl). 2. There are scarce possibilities for encountering insects or other wild animals (e.g., birds).	Any of the following conditions are met: 1. There are no pet animals at school. 2. There are scarce possibilities for encountering insects or other wild animals (e.g., birds).
Natural materials	Almost no presence of synthetic materials. Dominant presence of natural materials (e.g., timber finishes and furniture, stone, natural fibres, clay-based, unpainted bricks).	Intermediate situation between the other two options (e.g., natural materials are not dominant, some presence of non-synthetic materials, and natural materials painted with synthetic paints).	Almost no presence of natural materials. Great presence of synthetic materials (e.g., plastic furniture, melamine, synthetic fibres, synthetic carpets, and synthetic paint in doors and window frames).
Images of nature	All of the following conditions are met: 1. All children have access. 2. Diverse images (photos and/or drawings) showing multiple elements and human experiences. 3. Images created by children.	Some of the previous conditions are met.	None of the previous conditions are met.
Textures and patterns	All children have access to nature-like textures and patterns in furniture, flooring, walls, other architectural elements.	There are very few nature-like textures and patterns within the building or not all children have access.	Do not exist.

Table 2. *Cont.*

Attribute	Achieved	Partially Achieved	Not Achieved
Colour palette	Generally, colours are not strident and excessive (e.g., too many different contrasting and vibrant colours); though some bright colours in specific elements exist. Blues, green and/or earth tones stand out. Interior walls are light in colour (preferably white) and not blocked by excessive billboards. Exterior: green stands out.	There is a combination of colours between light walls, earth tones, blues and greens; but also, strident colours stand out.	Overall, the colour palette is not close to blues, green or earth tones. Only strident or grey colours stand out. Interior walls are not light in colour or absolutely blocked by excessive billboards and other elements.
Shapes and forms	All children have access to shapes and forms in the building elements such as: botanical motifs, tree and columnar supports, shells and spirals, oval and tubular forms, arches, vaults, domes, shapes resisting straight lines and right angles.	There are very few nature-like shapes and forms within the building, or not all children have access.	Not existing.
Views out	Both conditions are met: 1. Existing view of natural landscape, far away view (including broad urban views), natural elements (e.g., plants), or intense pedestrian traffic. 2. All children have views out from their seats.	Any of the following conditions are met: 1. Existing view of non-natural close elements or close street without pedestrians. 2. Existing view out but not from all children's seats or not at children's height. 3. Blinds are usually kept down.	Any of the following conditions are met: 1. No views out. 2. Windows to corridors, atriums or to other indoor room. 3. Windows above children's height.
Infrastructure for outdoor classes	There is infrastructure which allows outdoor classes, including shaded areas, space for seating, non-noisy environment, outdoor-boards or other outdoor-equipment.	There is little infrastructure but it is still possible to have classes outdoors.	There is no infrastructure (e.g., no shades or space to sit the whole class, and it is too noisy to work there).
Promotion of healthy commuting	There is enough space dedicated to parking bikes/scooters for more than half the number of children.	There is enough space dedicated to parking bikes/scooters for more than half the number of children.	There is no parking space for bikes/scooters.
Gardening sessions	Activity taking place on a regular basis for all children. The school has planters and/or gardening pots which children are allowed to interact with.	Activity taking place very sporadically or only for some children. The school has planters and/or gardening pots which children are allowed to interact with.	Not existing.
'Forest School' activities	There are 'Forest School' type activities on a regular basis for all children.	There are some 'Forest School' type activities seldom, very sporadically, and/or only for some children.	Do not exist.

Table 2. *Cont.*

Attribute	Achieved	Partially Achieved	Not Achieved
Spatial variability	At least three of these conditions are met: 1. Different accesses from the street to the school. 2. Several routes within the school or circular routes. 3. Winding routes/corridors. 4. Sequences of tight and broad spaces or diversity of circulation areas (e.g., mix between corridors and distribution halls).	Two or less of the previous conditions are met.	None of the previous conditions are met. There is a lack of spatial complexity (e.g., the access to different spaces is only through a common corridor, unique routes, straight corridors, and only one type of circulation).
Prospect	There are wide views beyond the limits of the school for all children.	There are some wide views within the limits of the school and/or only available for some children.	There are no wide views within the limits of the school or beyond.
Refuge	There are many possibilities for children to find a quiet and relatively out-of-sight spot while playing.	There are a few possibilities for children to find a quiet spot relatively out of sight while playing.	There are almost no possibilities for children to find a spot relatively out of sight while playing.
Active design	There are stairs that all kids use every day or there are: 1. Existing opportunities to walk and wander (e.g., spaces not too closely connected). 2. Steps, slopes and/or topographic variations in playgrounds.	There are stairs but not all children use them every day and/or there are not many opportunities for wandering or there are no topographical variations in the playgrounds.	There are no stairs, no opportunities for wandering (e.g., spaces too close together, all activities during the day take place in the same physical space), and no topographic variations on playgrounds.
Possibility to run	The playground is big enough to allow children to run, there are possibilities for children of different ages to have separated areas/playgrounds, and the paving surface is even and not too hard. There are not too many protruding elements around.	Some of the previous characteristics are not met.	There are no possibilities for children to run because the size of the playgrounds do not allow it or it is too dangerous (e.g., types of surfaces, protruding elements around) or because children of all ages have to share the same area at the same time.
Possibility to climb	The equipment in the playground (or sports room) allows a diversity of possibilities for climbing for all children.	The equipment in the playground (or sports room) allows some possibilities for climbing but not too much variation or they are not available for all children.	There are no possibilities (e.g., no playing equipment in playground).

2.3. *Case Studies and Audits*

Two case study primary schools were visited and audited against the new framework. The aim of the audits was to establish and compare the overall biophilic situation of the case studies, and to give background context to the data collected in these schools.

At the moment of the study, school A (London) had 300 (approx.) pupils, 18 teachers and teacher assistants and 23 other school staff. School B (Bath) had 210 (approx.) pupils, 22 teachers and teacher assistants and 7 other school staff.

2.4. Surveys

2.4.1. Teachers' Survey

The participants were current main teachers or assistants.

The call for participation was sent by email to the headteachers of the respective schools, who could opt for face-to-face, paper-based, or online-based formats. School A's headteacher handed paper-based versions at a teachers' weekly meeting and collected them two days later. Five responses were obtained from the estimated 15–18 teachers invited. The researcher was later invited to a summer fair where three more teachers participated face-to-face. School B's headteacher delivered paper-based questionnaires to the teachers after the school year ended and returned them to the researcher scanned by email. Six responses were collected from school B.

The questionnaire consisted of four open questions that aimed to evaluate the following:

1. Whether certain biophilic features have a perceived impact on children's preference, performance, and behaviour, through the lens of teachers.
2. What possible mechanisms could be causing the perceived impact.
3. Whether the school physical environment influences teachers' perception of the impact of the biophilic features on children.

Each question on the questionnaire targeted a different physical element from the school environment, associated to one or several biophilic attributes from the framework: "windows", "blinds up", "materials and finishes" and "plants in classrooms" (see Table 3). The questions asked about their perception of the effect of these conditions on children's preference, performance, and behaviour. Other aspects that arose, such as teachers' own preference, were also considered in the analysis.

Table 3. Questions from the teachers' questionnaire, their biophilic physical aspects associated and the relation to the framework.

Physical Aspect	Question as in the Questionnaire	Framework's Attributes Related to the Question
Windows	Q1: Do you notice any changes in performance or behaviour (attention or disruption) when children are seated next to the windows in the classroom?	Natural light, fresh air, and views out.
Blinds up	Q2: Do you usually have your window blinds down during class? How do children react to this? Do you notice a change in behaviour or performance when blinds are down or up, or when artificial light is on/off?	Natural light, views out.
Materials and finishes	Q3: Regarding the range of materials and finishes that could be found in the classroom, some may be more artificial or synthetic (e.g., plastic, melamine, plain colours, geometric patterns) and some might be natural or nature-like (e.g., timber, stone, clay based, natural fibres, nature-like patterns, textures and colours). Have you noticed any differences in children's reactions (preference, performance or behaviour) when interacting with this type of materials/finishes in comparison to artificial?	Natural materials, textures and patterns, and colour palette.
Plants in classroom	Q4: According to your experience, do you think more plants in a classroom could have a positive impact on children's performance and behaviour? Have you noticed changes in behaviour or performance when (if) plants were introduced in the classroom?	Plants

The analysis of this survey consisted of two parts: an overall coding analysis to determine whether the impacts were perceived, and if so whether these were positive or negative and a detailed qualitative analysis to collect the specific perceived effects on children's preference, performance and behaviour.

2.4.2. Parents' Survey

Parents were invited to participate through the parents' social media network and face to face in school social events. Parents from school A were invited in person during the school's summer fair. Parents from school B were invited through a parents Facebook group which included an introductory note written by one parent and a link to an online form served by SurveyMonkey.com. In total, 17 responses were collected from school A and 6 from school B.

The parents' questionnaire consisted of three checkbox questions aiming to capture the parents' perception on children's preferences, desires, and comments related to 8 attributes selected from the framework: natural light, water, plants, animals, views out, natural materials, images of nature, and colour palette (see Table 4). Question 1 includes options about non-physical aspects (activities and people, which are presumed to be easily mentioned by children) to confirm whether the children talk about their school environment at home.

Table 4. Questions from the parents' questionnaire and the relation to the framework. The options to be selected were accompanied by checkboxes.

Question as in the Questionnaire	Framework's Attributes Related to the Question
Q1: Do your children talk about their classroom environment at home (e.g., something they particularly like about their classroom)? Please select: Materials, colours, views, images, natural elements (plants, water features, animals), environmental conditions (daylight, temperature, smells), activities, classmates, teachers, other: ______________.	Natural light, water features, plants, animals, views out, natural materials, images of nature, and colour palette.
Q2: What do you think your children would like to change/add/remove in the classroom? Please select: Materials, colours, views, images, natural elements (plants, water features, animals), environmental conditions (daylight, temperature, smells), other: ______________.	Natural light, water features, plants, animals, views out, natural materials, images of nature, and colour palette.
Q3: Have your children ever mentioned they would like more natural elements in their school? Please select: Plants, water features, more daylight, trees, especial colours, images of nature or particular views, natural materials (timber, stone, clay, sand), other: ______________.	Water features, plants, natural materials, images of nature, and colour palette.

The analysis of this survey consisted of counting the sum of votes on each checkbox in order to

a. confirm whether children comment on school environments at home;
b. identify and compare which elements of biophilia are more perceived, commented and desired by the children;
c. identify differences on these between schools, and to analyse whether the school could have an impact on the type of elements mentioned and on the awareness of biophilia.

3. Results

3.1. Audits

Table 5 summarises the audits' results. Overall, in School A, 3 attributes were considered achieved (all representational), 11 partially achieved, and 7 were not achieved. Whereas for School B, 15 attributes were achieved, 4 partially achieved, and 2 were not achieved (both representational). Therefore, school A can be considered as a 'low-biophilic school' and school B a 'higher-biophilic school'.

Table 5. Summary of the audit results.

	Attribute	School A	School B
Direct experience (presence of elements)	Natural light	🟡	🟢
	Fresh air	🟡	🟡
	Water	🟡	🟢
	Plants	🟤	🟢
	Animals	🟡	🟢
	Natural materials	🟤	🟢
Direct experience (spatial interactions)	Views out	🟡	🟢
	Infrastructure for outdoor classes	🟤	🟢
	Promotion of healthy commuting	🟡	🟡
	Gardening sessions	🟤	🟢
	'Forest School' activities	🟤	🟢
Representation (presence of elements)	Images of nature	🟢	🟢
	Textures and patterns	🟤	🟤
	Colour palette	🟡	🟢
	Shapes and forms	🟡	🟤
Representation (spatial interactions)	Spatial variability	🟡	🟢
	Prospect	🟡	🟡
	Refuge	🟤	🟢
	Active design	🟢	🟢
	Possibility to run	🟢	🟡
	Possibility to climb	🟡	🟢

3.2. Teachers' Surveys

Overall, the results from the survey show the following:

- Teachers from school B had a more unanimous and generally positive opinion about the impact of the biophilic features on children's preference, performance and behaviour than teachers from school A.
- The most conflicting responses were regarding windows and blinds, which were perceived positive by school B teachers and generally not well rated from school A. While teachers from school A used words such as "distraction", "disruption", or "I don't think they notice the blinds are down"; school B teachers used "enjoy", "preferred", "fascinated", "engaged", and "wake their brains up".
- Teachers from school B did not express the perception of negative impact for any condition.
- The most positively rated conditions were "plants in classrooms" for school A, and "plants in classrooms" and "windows" for school B.
- "Materials and finishes" showed greater uncertainty, but interestingly, some teachers commented that children are more respectful with objects and toys made of wood than plastic, which they throw away more easily, considering them disposable.

Figure 1 summarises the teachers' responses decoded. The teachers' responses (T) are expressed in the rows. The different questions (Q, columns) are subdivided into columns, each column referring to the different aspects of wellbeing analysed (teacher's preference, perceived children's preference, perceived children's performance, and perceived children's behaviour). For each question, the coloured icons rating the perceived impacts were allocated under the corresponding wellbeing aspect they were referring to. E.g., the quote of T1 for Q1 who says: "More disruption as they often look out and stare which means we have to stop the lesson and remind them to focus" was decoded as (a) "more disruption": perceived negative impact on children's behaviour; (b) "remind them to focus": perceived negative impact on children performance; and (c) overall: perceived children's positive preference. So, despite the question referring to children's performance and behaviour, positive children's preference was manifested too, marked with a green icon under "children's preference", and red icons under "performance" and "behaviour".

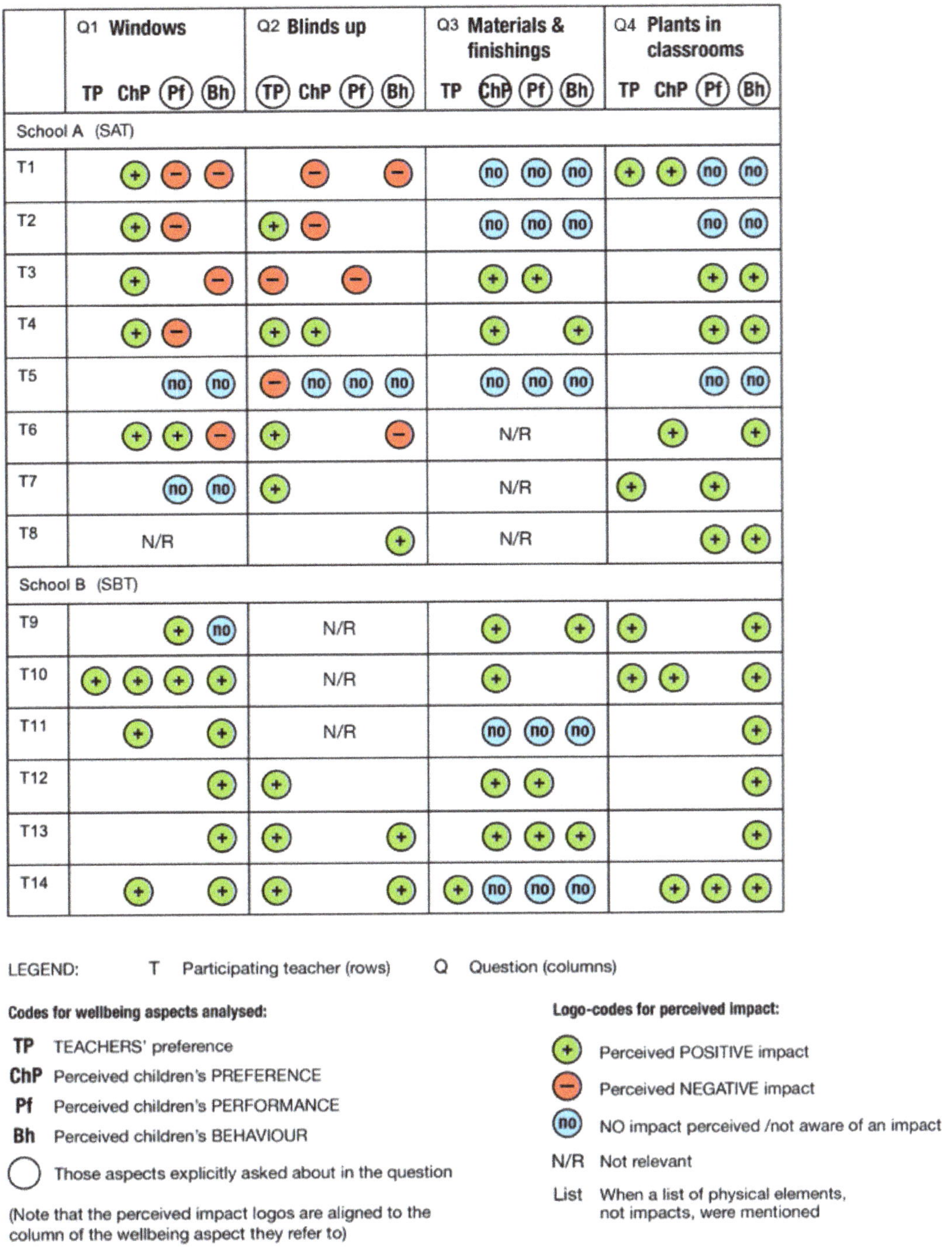

Figure 1. Comparative analysis of teachers' survey.

3.3. Parents' Surveys

When asked about whether their children talk about their classroom environment at home (Figure 2, Table 6), the most popular responses were related to their classmates, teachers and activities. This confirms the fact that children actually speak about their school environment at home. From the other options, school A parents mentioned aspects not necessarily related to biophilia, whereas school B parents answered yes, and in a larger proportion.

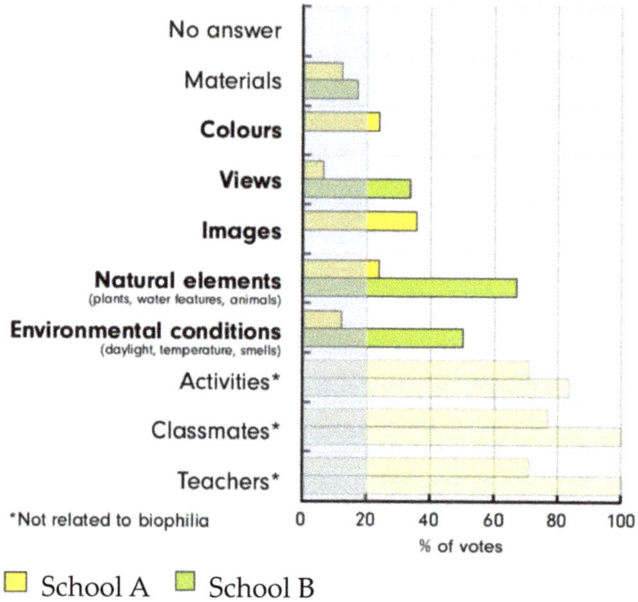

Figure 2. Results from parents' question 1. The grey shade shows answers below a 20% response.

Table 6. Results from parents' question 1: "Do your children talk about their classroom environment at home? Do they mention any of the following?".

	School A (n = 17)		School B (n = 6)	
	Votes	**%**	**Votes**	**%**
No answer	0	0	0	0
Materials	2	12	1	17
Colours	4	24	0	0
Views	1	6	2	33
Images	6	35	0	0
Natural elements	4	24	4	67
Environmental conditions	2	12	3	50
Activities	12	71	5	83
Classmates	13	77	6	100
Teachers	12	71	6	100

- "Images", "colours", and "natural elements" were the top three options that parents from school A selected (35%, 24% and 24% of parents respectively).
- For school B, the top three options were "natural elements", "environmental conditions", and "views" (67%, 50%, and 33% respectively).
- For school A, the features related to the representation of nature were the most popular. Whereas for school B, the most popular options selected relate to the direct experience of nature, and nobody selected "colours" or "images".
- Relevant statements at the option "other" were "Food" and "The children do not tend to talk about the fixed features—more on people".

For question 2, "what do you think your children would like to change, add or remove in their classroom?" (Figure 3, Table 7), a large proportion of parents from school A did not answer the question (41%), whereas all parents from school B responded. The outstanding responses for both groups were "natural elements" (83% of parents from school B and 18% from school A) and "environmental conditions" (67% for school B and 18% for school A).

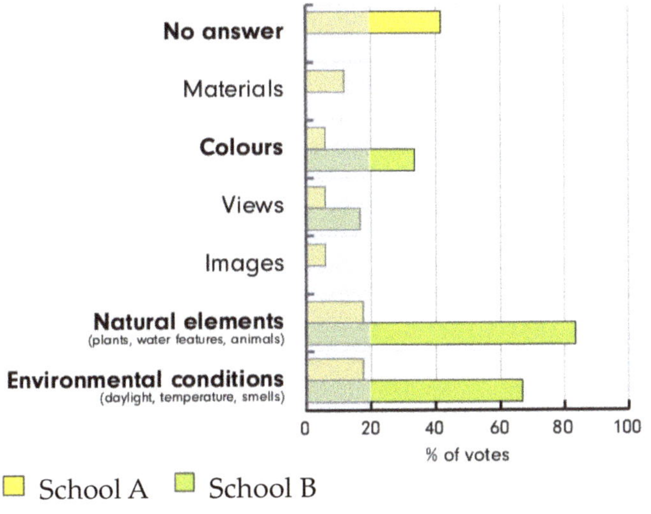

Figure 3. Results from parents' question 2. The grey shade hides those answer with less than a 20% response.

Table 7. Results from parents' question 2: "What do you think your children would like to change, add or remove in their classroom?".

	School A (n = 17)		School B (n = 6)	
	Votes	**%**	**Votes**	**%**
No answer	7	41	0	0
Materials	2	12	0	0
Colours	1	6	2	33
Views	1	6	1	17
Images	1	6	0	0
Natural elements	3	18	5	83
Environmental conditions	3	18	4	67

Relevant statements at the option "other" were "If they had the option to provide views, I am sure they would engage", "loves water & aquatic plants & animals, and looking after his plants at home".

When parents were explicitly asked about natural elements (Figure 4, Table 8), the most popular answer was "plants" (24% response rate for school A and 83% for school B). Especially in school B where they actually have plants everywhere in the school. Again, for this question, most parents from school A left it empty and the rest of the options were smoothly distributed. The second most selected options for parents from school B were "water features", "trees" and "natural materials" (33% each). The only option not selected by school B parents was "images of nature or particular views".

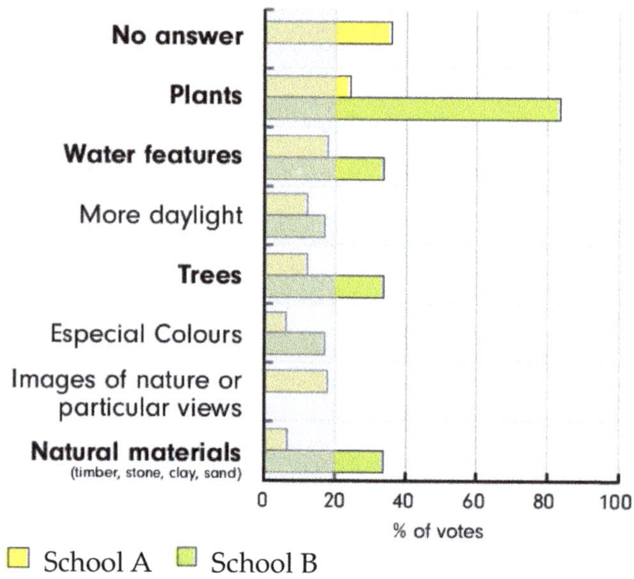

Figure 4. Results from parents' question 3. The grey shade hides those answer with less than a 20% response.

Table 8. Results from parents' question 3: "Have your children ever mentioned they would like more natural elements in their school?".

	School A (n = 17)		School B (n = 6)	
	Votes	**%**	**Votes**	**%**
No answer	6	35	0	0
Plants	4	24	5	83
Water features	3	18	2	33
More daylight	2	12	1	17
Trees	2	12	2	33
Especial colours	1	6	1	17
Images of nature or particular views	3	18	0	0
Natural materials	1	6	2	33

Relevant statements in the option "other" for this question were "an indoor waterfall", "fish" and "to see the sky".

4. Discussion

4.1. Overall Comparison of Results

A framework to evaluate primary school environments based on their success in implementing biophilic design (objective 1) was developed, based on the literature review and its adaptation to school environments. Performance and behaviour (objective 2) were analysed through the teachers' survey and preferences (objective 3) through the surveys to teachers and parents.

4.1.1. In Relation to Objective 2: Teachers' Perception of Certain Biophilic Features on Children's Performance and Behaviour

Overall, results suggest that the biophilic features analysed have a perceived impact on children's performance and behaviour, and that there are differences from "low-biophilic" and "higher-biophilic" schools (represented by school A and school B, respectively).

Performance:

- According to teachers, children have a preference towards windows, which leads to some types of distraction. Therefore, windows were negatively qualified by some teachers from school A, but considered very positive by teachers from school B.
- Views out, natural materials and plants were perceived as promoting observations, curiosity and discussions.

Behaviour:

- Teachers from both schools agree that children behave carefully and respectfully towards plants and objects made of natural materials (i.e., wood). Also, they agree that plants are calming for the children.

4.1.2. In Relation to Objective 3: Children's Preference towards Biophilic Features, as Perceived by Teachers and Parents

According to parents, children comment more on activities and people than on physical characteristics of the physical environment. A parent's comment at the survey summarises it: "Children do not tend to talk about the fixed features, more on people".

However, though to a lesser extent, children do comment on the school's physical characteristics. Contrasting parents' perspectives, children from school A mention biophilic features much less often than children from school B. Besides, children from school B comment more on features classified in the framework as "direct experience" (i.e., natural elements, environmental conditions, and views), a category in which this school was highly valued in at the audit; while students from school A comment more on features from the "indirect experience" category (i.e., images and colours).

Regarding children's desire, 35% of parents from school A do not know what children would want to change or desire at their school environment, while all parents from school B had an answer. Although the audit result considered that school B had a higher biophilic condition compared to school A, 83% of parents from school B expressed that their children would like more plants, while only 24% of parents from school A did.

4.2. Key Findings

4.2.1. Teachers' Awareness

On some aspects, teachers from the "higher-biophilic" school perceive more benefits and seem more sensitive to the school's environment conditions and their effect on children than teachers from the "low-biophilic" school. For instance, the former expressed awareness on the benefits of daylight, views out and fresh air, whereas the latter manifested little awareness and made comments such as "I don't think they notice" regarding whether the blinds were down or up.

4.2.2. Children's Awareness and Demand

Results suggest that children's preference for biophilic features is stronger when the exposure is higher. Hypothetical reasons could be:

a. The richer the environment, the higher their understanding, awareness of biophilic features and desire for other features and expectations.
b. The role that teachers play in stimulating awareness. Teachers could be more sensitive too, due to the environment where they work.

4.2.3. Design Effectiveness vs. Impact of the Actual Element

Teachers' perception of the impact of biophilic features on children's wellbeing could be restricted by inadequate building design (e.g., lack of windows shading, leading to glare). In this context, "blinds up" were perceived as undesired and perturbing the general performance of children, when actually glare might have been the cause. On the other hand, parents expressed that children do want more daylight and teachers themselves believe children do have a preference for windows. Another possible justification for the reported misbehaviour could be that children without enough exposure to certain biophilic features, in need of restoration or biophilic fulfilment "get distracted" with the views out and daylight because

a. they do not usually have them,
b. their biophilic needs are not satisfied,
c. they associate these biophilic features with the playground environment, space where they experience movement, restoration, and where they have more exposure to biophilic features.

4.2.4. Role of Teachers

The results from the teachers' surveys evidence their fundamental role in the control of the environmental conditions and on the effectiveness of the biophilic attributes through encouragement of observation and the activities performed.

The previous research findings are tentative, since the samples were limited. Further research should tackle these key points to confirm or reject the hypothesis formulated.

4.3. Findings in Relation to the Literature

The findings in this study support Lee and Park's [1] principles for space design in children's libraries based on biophilic design patterns, extending them to the wider school environment (classroom and patio). In particular, the offer of the experience of nature, shelter and open space, and a multisensory approach. This study also evidences the role of "fascination (cognitive or physical)" in restorativeness described by Bagot, Allen and Toukhsati [37] and the positive influence of natural elements in students' performance as described by Han [43]. The literature also suggests that a lack of natural experiences could lead to a lesser care of the natural environment [3,38,51–54], which could somehow be reflected in this study regarding objects made out of natural materials vs. synthetic, since the former are perceived to be more respected. This could also be due to the extended use of disposable plastic that children encounter every day.

Han [43] also emphasises the importance of a learning environment which reduces stress and advocates for children to concentrate; it is evident in this study that an environment of discomfort, such as when glare is present, can lead to misbehaviour. A possible reason for children's reported misbehaviour in the presence of daylight could simply be the impossibility to relocate (i.e., move away) when glare perturbs them. Parpairi [55] found that visual satisfaction of students in Cambridge libraries was not associated to the actual measured light levels and was high even under uncomfortable situations (e.g., glare) if there were good views of landscape and the person had the option to move. Adaptation theories, referring to thermal comfort [56,57], could have a role here too. Baker [58] depicts this situation in his "general theory of environmental comfort" as the relationship between adaptive opportunities and stress, highlighting the importance of providing relocating opportunities to users. In the case of school A, not having the option of relocation while being affected by glare might be the cause for children's misbehaviour.

Lastly, the data gathered provided some new elements of nature to consider, which had not been covered in the existing biophilia frameworks, such as sky views and shades.

4.4. Contributions, Limitations, and Future Research Directions

This study is relevant to different disciplines including designers, teachers, school authorities, and policy makers. Below there are some reflections that can be of interest to those keen to conduct similar studies to confirm the findings.

4.4.1. About the Proposed Framework

This study developed a new biophilic framework of analysis which focuses on features specifically applicable to primary school environments, and which allows a basic evaluation of the school's biophilic conditions under a fulfilment criterion. However, it needs further testing in a diversity of school environments and with a diverse range of auditors. A quantifiable set of criteria for evaluating the biophilic attributes would benefit this framework. Further research could target ways to quantify the dose and length of exposure of the different features, and the repercussions that these will have on the impact on children. This could include minimum areas of vision, times for exposure, and percentage rates, etc.

4.4.2. Methodological Considerations

- Sample: The samples (schools and participants) were limited. A larger number of schools and a broader diversity of biophilic conditions could enrich the results.
- Demographic information: Questionnaires in this study were kept short to reduce unresponsiveness; therefore, demographic information was not collected. Gathering information related to gender, age, and socioeconomic context could capture aspects that could impact the results as well.
- Asking children: Nobody better than children could know what they prefer or desire. This research suggests that (a) parents are not always sure about children's desire towards their school's physical environment and (b) teachers might confuse children's desire and biophilic needs with discomfort issues caused by inadequate design. Teachers' opinions about the impacts on children wellbeing varies as well (e.g., several teachers consider that views out have a strong impact, while others believe children do not realise whether blinds are down or up). As Bagot, Allen and Toukhsati say: "The inclusion of children in the understanding of their environments is warranted and is likely to give rise to elements not considered by adults thus far" [37] (p. 8).
- Experimental studies: Future research could use experimental variables, such as controlling blinds, exposing children to different types of views out, different lighting and glare conditions, options for relocation, testing performance, observing children behaviour and surveying teachers and children. Experimental studies with vulnerable populations (children) would require stricter ethical considerations but would provide very valuable information.

4.4.3. Other Considerations

- Location and socio-economic factors: Future research with a larger sample of schools could investigate whether some factors such as city size, socio-economic factors related to location, proximity to green areas or natural landscapes, and/or the biophilic condition of the city itself have a relevant impact on the results.
- Role of parents: This research focused on the biophilic conditions of school environments, since it is here where children spend most of their time after home; but further information could have been identified on the time at home and on weekends, in terms of whether children ever experienced "direct nature" (in the sense of Louv [38]). While

in this study the role of teachers in encouraging the observation of biophilic features for children was considered, the role of the parents was not. Research is needed on the parents' own relation to nature and how much this influences the school selection.

4.4.4. Implications for Practice and Policy

- For designers: This research exposes the impact that inadequate design and discomfort can have on teachers' and children's perception of biophilic features. It also highlights the relevance of providing opportunities for restoration and changes of focus, not only in playgrounds but also in classrooms. The framework proposed in this study can be used as a tool to support the design process.
- For teachers and headteachers: the study highlights the significant role of teachers in managing classroom environmental conditions (e.g., use of blinds) and in enhancing biophilia features/awareness through the encouragement of observation and promotion of discussion. Recommendations include incrementing the opportunities of being outside (e.g., outdoor curricular activities) and including more biophilic features in classrooms (e.g., wooden furniture).
- For policy: Biophilic design is suggested to be an essential consideration for the correct development of children's wellbeing. This implies not only infrastructure (e.g., proper space for outdoor classes and required equipment, recondition of playgrounds, and use of natural materials) but also teachers' encouragement of biophilic activities (e.g., gardening, active commuting).

5. Conclusions

This mixed-method study focusing on teachers' and parents' views suggests that the school's physical environment could impact children's preference, performance and behaviour. Results revealed that a higher biophilic environment increases teachers' and children's sensitivity, awareness and demand towards biophilic features themselves. Furthermore, teachers who were exposed to biophilic features had a positive perception of their effects. Similarly, teachers and parents from schools where children are more exposed to biophilic elements perceived that their children are more aware and comment more on these types of elements than children who are less exposed. Additionally, inadequate design of the building (e.g., insufficient windows shading), which may lead to discomfort (e.g., glare), could have an effect on the perception of negative impact of biophilic attributes (e.g., daylight, views out). This study, therefore, has implications for design and for educational policies.

This study's contribution to the research field also includes a bespoke framework for biophilic design in primary school environments which allows biophilia-oriented evaluations of existing school buildings or could be used as a tool to aid the design stage. Furthermore, the combination of audits and surveys can enhance the understanding of how the perception of biophilic features could be conditioned by the physical environment, and how this could impact children's preference, performance and behaviour.

Further research should examine the reasons that determine children's preferences and demands for biophilic features when exposed to higher or lower biophilic environments (e.g., better understanding, expectations, and role of adults). The role of parents on the promotion of biophilia, their own biophilic preferences and any relation to biophilia when choosing their children's school could be considered too.

Author Contributions: Conceptualization, B.L. and M.U.; methodology, B.L.; validation, M.U.; formal analysis, B.L.; investigation, B.L.; writing—original draft preparation, B.L.; writing—review and editing, M.U.; visualization, B.L.; supervision, M.U. All authors have read and agreed to the published version of the manuscript.

Funding: This research received no external funding.

Institutional Review Board Statement: All the information gathered by this research was anonymous (no personal data) and participation absolutely voluntary. The research received Low-risk ethical approval from BSEER, UCL. In order to be applicable for Low-risk approval, participation excluded pupils.

Informed Consent Statement: Informed consent was obtained from all subjects involved in the study.

Data Availability Statement: The data of the surveys presented in this study are available on request from the corresponding author, who can further explain relevant sections where needed. Photography that supports the audits presented are not available since authorization was not provided.

Acknowledgments: Alba Fernandez Arias, Jennifer Lorrimar-Shanks, and Delia Masanotti, for providing links to the case study schools; the case study primary schools and respective headteachers; and all the participants involved.

Conflicts of Interest: The authors declare no conflict of interest.

References

1. Lee, H.C.; Park, S.J. Assessment of importance and characteristics of biophilic design patterns in a children's library. *Sustainability* **2018**, *10*, 987. [CrossRef]
2. Heerwagen, J.H.; Orians, G.H. Humans, Habitats, and Aesthetics. In *The Biophilia Hypothesis*; Kellert, S.R., Wilson, E.O., Eds.; Island Press: Washington, DC, USA, 1993; pp. 138–172.
3. Ulrich, R.S. Biophilia, Biophobia and Natural Landscapes. In *The Biophilia Hypothesis*; Kellert, S.R., Wilson, E.O., Eds.; Island Press: Washington, DC, USA, 1993; pp. 73–137.
4. Children's Environmental Health—Environmental Risks, World Health Organization. Available online: https://www.who.int/ceh/risks/en/ (accessed on 19 July 2019).
5. Plotka, E. *Better Spaces for Learning*; Royal Institute of British Architects: London, UK, 2016.
6. Conners, D.A. The School Environment: A Link to Understanding Stress. *Theory Pract.* **1983**, *22*, 15–20. [CrossRef]
7. Kopec, D.A. Schools and Learning Environments. In *Environmental Psychology for Design*; Fairchild: New York, NY, USA, 2006; pp. 188–209.
8. Barrett, P.; Treves, A.; Shmis, T.; Ambasz, D. *The Impact of School Infrastructure on Learning: A Synthesis of the Evidence*; World Bank: Washington, DC, USA, 2019. [CrossRef]
9. Kellert, S.R. *Nature by Design*; Yale University Press: New Heaven, CT, USA, 2018.
10. Appleton, J. *The Experience of Landscape*; Wiley: London, UK, 1975.
11. Larsen, L.; Adams, J.; Deal, B.; Kweon, B.S.; Tyler, E. Plants in the Workplace: The Effects of Plant Density on Productivity, Attitudes and Perceptions. *Environ. Behav.* **1998**, *30*, 261–281. [CrossRef]
12. Nakamura, R.; Fujii, E. Studies of the Characteristics of the Electroencephalogram when Observing Potted Plants: Pelargonium Hortorum "Spriner Red" and Begonia Evansiana. *Tech. Bull. Fac. Hortic. Chiba Univ.* **1990**, *43*, 177–183.
13. Lohr, V.I.; Pearson-Mims, C.H.; Goodwin, G.K. Interior plants may improve worker productivity and reduce stress in a windowless environment. *J. Environ. Hortic.* **1996**, *14*, 97–199. [CrossRef]
14. Chang, C.Y.; Chen, P.K. Human response to window views and indoor plants in the workplace. *HortScience* **2005**, *40*, 1354–1359. [CrossRef]
15. Orsega-Smith, E.; Mowen, A.J.; Payne, L.L.; Godbey, G. The Interaction of Stress and Park Use on Psycho-physiological Health in Older Adults. *J. Leis. Res.* **2018**, *36*, 232–256. [CrossRef]
16. Hung, M.F.; Chang, C.Y. The effect of plants on preschool children's attention in the classroom. *Hortic. NCHU* **2002**, *27*, 77–86.
17. Shibata, S.; Suzuki, N. Effects of the foliage plant on task performance and mood. *J. Environ. Psychol.* **2002**, *22*, 265–272. [CrossRef]
18. Kuo, F.E.; Faber Taylor, A. A Potential Natural Treatment for Attention-Deficit/Hyperactivity Disorder: Evidence from a National Study. *Am. J. Public Health* **2004**, *94*, 1580–1587. [CrossRef]
19. Ottosson, J.; Grahn, P. A comparison of leisure time spent in a garden with leisure time spent indoors: On measures of restoration in residents in geriatric care. *Landsc. Res.* **2005**, *30*, 23–55. [CrossRef]
20. Mooney, P.; Nicell, P.L. The Importance of Exterior Environment for Alzheimer Residents: Effective Care and Risk Management. *Healthc. Manag. Forum. Can. Coll. Health Serv. Exec.* **1992**, *5*, 23–29. [CrossRef]
21. Taylor, A.F.; Wiley, A.; Kuo, F.E.; Sullivan, W.C. Growing up in the Inner City. *Environ. Behav.* **1998**, *30*, 3–27. [CrossRef]
22. Kuo, F.E.; Sullivan, W.C. Aggression and Violence in the Inner City. *Environ. Behav.* **2003**, *33*, 543–571. [CrossRef]
23. Moore, E.O. Prison Environment's Effect on Health Care Service Demands. *J. Environ. Syst.* **1981**, *11*, 17–34. [CrossRef]
24. Ulrich, R.S. View through a Window May Influence Recovery from Surgery. *Science* **1984**, *224*, 420–421. [CrossRef]
25. Fjeld, T.; Veiersted, B.; Sandvik, L.; Riise, G.; Levy, F. The Effect of Indoor Foliage Plants on Health and Discomfort Symptoms among Office Workers. *Indoor Built Environ.* **1988**, *7*, 204–209. [CrossRef]
26. Takano, T.; Nakamura, K.; Watanabe, M. Urban residential environments and senior citizens' longevity in megacity areas: The importance of walkable green spaces. *J. Epidemiol. Community Health* **2002**, *56*, 913–918. [CrossRef]

27. Rice, C.S.; Torquati, J.C. Assessing Connections between Young Children's Affinity for Nature and Their Experiences in Natural Outdoor Settings in Preschools. *Child. Youth Environ.* **2013**, *23*, 78–102. [CrossRef]
28. Maller, C.J. Promoting children's mental, emotional and social health through contact with nature: A model. *Health Educ.* **2009**, *109*, 522–543. [CrossRef]
29. Kaplan, R.; Kaplan, S. *The Experience of Nature: A Psychological Perspective*; Ulrich's Bookstore: Ann Arbor, MI, USA, 1995.
30. Hartig, T. Restorative Environments. *Encycl. Appl. Psychol.* **2004**, *3*, 273–279. [CrossRef]
31. Berman, M.G.; Jonides, J.; Kaplan, S. The Cognitive Benefits of Interacting with Nature | The Center for Health Design. *Psychol. Sci. Res.* **2008**, *19*, 1207–1212. [CrossRef]
32. Stevenson, M.P.; Dewhurst, R.; Schilhab, T.; Bentsen, P. Cognitive restoration in children following exposure to nature: Evidence from the attention network task and mobile eye tracking. *Front. Psychol.* **2019**, *10*, 429820. [CrossRef]
33. Verderber, S. Dimension of Person-Window Transactions in the Hospital Environment. *Environ. Behav.* **1986**, *18*, 450–466. [CrossRef]
34. Malenbaum, S.; Keefe, F.J.; Williams, A.C.D.C.; Ulrich, R.; Somers, T.J. Pain in its environmental context: Implications for designing environments to enhance pain control. *Pain* **2008**, *134*, 241–244. [CrossRef]
35. Ulrich, R.S.; Zimring, C.; Zhu, X.; DuBose, J.; Seo, H.B.; Choi, Y.S.; Quan, X.; Joseph, A. A Review of the Research Literature on Evidence-Based Healthcare Design (Part I) Abstract Introduction Background. *Health Environ. Res. Des. J.* **2008**, *1*, 61–125. [CrossRef]
36. Andrade, C.C.; Devlin, A.S.; Pereira, C.R.; Lima, M.L. Do the hospital rooms make a difference for patients' stress? A multilevel analysis of the role of perceived control, positive distraction, and social support. *J. Environ. Psychol.* **2017**, *53*, 63–72. [CrossRef]
37. Bagot, K.L.; Allen, F.C.L.; Toukhsati, S. Perceived restorativeness of children's school playground environments: Nature, playground features and play period experiences. *J. Environ. Psychol.* **2015**, *41*, 1–9. [CrossRef]
38. Louv, R. *Last Child in the Woods: Saving Our Children from Nature-Deficit Disorder*; Algonquin Books of Chapel Hill: Chapel Hill, NC, USA, 2006.
39. Chawla, L.; Keena, K.; Pevec, I.; Stanley, E. Green schoolyards as havens from stress and resources for resilience in childhood and adolescence. *Health Place* **2014**, *28*, 1–13. [CrossRef]
40. Hodson, C.B.; Sander, H.A. Green urban landscapes and school-level academic performance. *Landsc. Urban Plan.* **2016**, *160*, 16–27. [CrossRef]
41. Benfield, J.A.; Rainbolt, G.N.; Bell, P.A.; Donovan, G.H. Classrooms with Nature Views: Evidence of Differing Student Perceptions and Behaviors. *Environ. Behav.* **2015**, *47*, 140–157. [CrossRef]
42. Li, D.; Sullivan, W.C. Impact of views to school landscapes on recovery from stress and mental fatigue. *Landsc. Urban Plan.* **2016**, *148*, 149–158. [CrossRef]
43. Han, K.T. Influence of Limitedly Visible Leafy Indoor Plants on the Psychology, Behavior, and Health of Students at a Junior High School in Taiwan. *Indoor Built Environ.* **2009**, *4*, 658–692. [CrossRef]
44. Oseland, N. The impact of psychological needs on office design. *J. Corp. Real Estate* **2009**, *11*, 244–254. [CrossRef]
45. Kellert, S.R. Experiencing nature: Affective, Cognitive and Evaluative Development in Children. In *Children and Nature: Psychological, Sociocultural, and Evolutionary Investigations*; Kahn, P.H.J., Kellert, S.R., Eds.; The MIT Press: Cambridge, MA, USA; London, UK, 2002.
46. Kellert, S.R. Dimensions, Elements and Attributes of Biophilic Design. In *Biophilic Design: The Theory, Science, and Practice of Bringing Buildings to Life*; Kellert, S.R., Heerwagen, J.H., Mador, M.L., Eds.; Wiley: Hoboken, NJ, USA, 2008; pp. 3–19.
47. Heerwagen, J.H.; Gregory, B. Biophilia and Sensory Aesthetics. In *Biophilic Design: The Theory, Science, and Practice of Bringing Buildings to Life*; Kellert, S.R., Heerwagen, J.H., Mador, M.L., Eds.; Wiley: Hoboken, NJ, USA, 2008; pp. 227–241.
48. Browing, W.; Ryan, C.; Clancy, J. *14 Patterns of Biophilic Design: Improving Health & Well-Being in the Built Environment*; Terrapin Bright Green: New York, NY, USA, 2014.
49. Biophilic Design Initiative. Available online: https://living-future.org/biophilic-design/ (accessed on 26 February 2024).
50. WELL Building Standard v2: Concepts and Features, International WELL Buildings Institute. Available online: https://dev-wellv2.wellcertified.com/v/en/concepts (accessed on 22 August 2019).
51. Kellert, S.R.; Wilson, E.O. *The Biophilia Hypothesis*; Island Press: Washington, DC, USA, 1993.
52. Zhang, W.; Goodale, E.; Chen, J. How contact with nature affects children's biophilia, biophobia and conservation attitude in China. *Biol. Conserv.* **2014**, *177*, 109–116. [CrossRef]
53. Mustapa, N.D.; Maliki, N.Z.; Hamzah, A. Repositioning Children's Developmental Needs in Space Planning: A Review of Connection to Nature. *Procedia Soc. Behav. Sci.* **2015**, *170*, 330–339. [CrossRef]
54. Preuß, M.; Nieuwenhuijsen, M.; Marquez, S.; Cirach, M.; Dadvand, P.; Triguero-Mas, M.; Gidlow, C.; Grazuleviciene, R.; Kruize, H.; Zijlema, W. Low Childhood Nature Exposure is Associated with Worse Mental Health in Adulthood. *Int. J. Environ. Res. Public Health* **2019**, *16*, 1809. [CrossRef]
55. Parpairi, K. *Daylighting in Architecture: Quality and User Preferences*; University of Cambridge: Cambridge, UK, 1999.
56. Baker, N. We are all outdoor animals. In *Architecture, City, Environment, Proceedings of PLEA 2000*; Steemers, K., Yannas, S., Eds.; James & James: London, UK, 2000; pp. 553–555.

57. Humphreys, M.; Nicol, F.; Roaf, S. *Adaptive Thermal Comfort: Foundations and Analysis*; Routledge: Abingdon, UK; New York, NY, USA, 2016.
58. Baker, N. Cultural Responses to Primitive Needs. In *Beyond Environmental Comfort*; Ong, B.L., Ed.; Routledge: New York, NY, USA, 2013; pp. 47–59.

architecture

MDPI

Article

A Toolkit of Biophilic Interventions for Existing Schools to Enhance Student and Faculty Health and Performance

Kari Leif * and Vivian Loftness

School of Architecture, Carnegie Mellon University, Pittsburgh, PA 15213, USA; loftness@cmu.edu
* Correspondence: kleif@andrew.cmu.edu

Abstract: School learning environments play a crucial role in both student and faculty outcomes; however, the limited funding allocated to public school facilities can result in poor environmental conditions which can hinder occupant health and performance. Existing school facilities must then explore affordable retrofit strategies that can effectively improve health and performance outcomes. The emerging field of biophilic design offers significant potential for improving existing school environments with benefits for both the students and faculty. Through case study research, this study proposes a toolkit of 42 biophilic retrofits for existing K-12 schools in a set of stakeholder cards that illustrate precedents, known impacts, and their relevance to high-performance schools. Additionally, a stakeholder card sorting study was conducted to establish the perceived viability and impact of each strategy. The findings reveal that biophilic retrofit design strategies are perceived by school community stakeholders to be impactful with varying levels of affordability. These findings further demonstrate that a toolkit of biophilic interventions for K-12 schools will offer invaluable insights to improve student and faculty conditions.

Keywords: biophilic design; K-12 schools; occupant health and performance; healthy buildings

1. Introduction

School learning environments play a crucial role in the development and wellbeing of children. However, research indicates that an alarming number of school facilities, especially those in public schools, face a funding gap that affects the quality of the school infrastructure [1–4]. K-12 students in both public and private schools spend an average of 42% of their waking hours within school facilities, thus it is imperative that school buildings are in excellent condition to enhance student learning [5]. Recent research has identified that quality school building design and maintenance have measurable effects on both student and teacher performance [6]. However, the American Society of Civil Engineers' (ASCE) 2021 Report Card for America's Infrastructure scored school facility conditions and building systems at a grade of D+ on an A–F scale in terms of structure and maintenance [7]. Given that both students and teachers spend a considerable amount of their daily lives in school buildings, facility design and retrofits need to actively enhance human health, wellbeing, and performance. In order to combat poor facility conditions, there needs to be a push for both the budgetary decisions to allocate increased funding to K-12 schools, as well as a push for community decision makers to invest in affordable retrofit options that have significant benefits to the health and performance of students and teachers.

Biophilia is the innate human desire to connect with nature and other living beings. The term "biophilia" was first coined by psychoanalyst Erich Fromm in his 1964 publication, *The Anatomy of Human Destructiveness* [8]. Fromm stated that biophilia was the "passionate love of life and all that is alive..." [8]. In 1984, biologist EO Wilson popularized the term "biophilia" and proposed the current definition for biophilia—humans' innate tendency to focus on living things [9]. However, it was not until 2008 that Stephen Kellert solidified the concept of biophilia as a strategy for architecture and sustainability and coined the term "biophilic design" [10]. Over the past 50 years, the term "biophilia" has shifted from a psychological concept to a design practice within the built environment.

Biophilic design is the practice of connecting people and nature within our built environments and communities. There are a growing number of resources to help facilitate biophilic design such as taxonomies and best practices. Two leading biophilic design guidelines are Stephen Kellert's Six Biophilic Design Elements [10] and Terrapin Bright Green's 14 Patterns of Biophilic Design [11].

In parallel, there is a growing body of literature that reveals the benefits of connecting children with nature, as well as the importance of biophilic design for human outcomes [12–20]. This research suggests that K-12 school decision makers should be investing in biophilic retrofits for schools because nature is key to early childhood development. In a 2005 publication, Kellert concluded that the healthy emotional, evaluative, and intellectual development of children depends on accessible and abundant natural environments [21]. Additionally, in a 2021 population-based birth cohort study in Metro Vancouver, Canada, Jarvis et al. identified that every 10% increase in percentage of vegetation within 250 m of a residential postal code resulted in a 0.16 increase in teacher-assessed Early Development Instrument scores [22].

Indeed, the concepts of biophilia and biophilic design are quite prevalent in the literature and have been the subject of over 700 published articles and studies as found in a 2021 critical review [23]. However, there is still a lack of widespread realization of biophilic design in architecture and the built environment. This study aims to continue bridging the gap between academia and practice through the creation and validation of a biophilic design toolkit for K-12 school community stakeholders.

After an extensive literature and design precedent review, seven categories of biophilic retrofit design strategies were captured in a biophilic retrofit taxonomy. These overarching categories include: Indoor Greenery, Views, Biophilic Finishes, Natural Light, Nature's Sounds, Taste and Touch, Nature Interaction and Engagement, and Spatial Biophilic approaches. A total of 42 biophilic design strategies are summarized in the taxonomy card set. In order to quantify the connection between the biophilic design strategies and their impact on student and teacher performance, a thorough literature review identifying the relevant benefits of specific biophilic retrofit strategies to humans was completed and captured on the back of each strategy card. The studies found significant connections between biophilic design elements and human health and productivity outcomes, including anxiety and stress [24–26], attention [27–30], cognitive function [12,13,19,31], engagement (in class) [32], graduation rates [33], physical health [15,34–37], standardized test scores [18,38–41], systems thinking [42], teamwork [43], and thermal comfort [44].

Through the understanding of biophilic design and its impact on humans, a rich palette of biophilic design elements can be developed to engage K-12 school community stakeholders and address the health and performance of both students and faculty.

2. Materials and Methodology

The Biophilic Toolkit and user testing were undertaken in two stages. The first stage focused on the creation of a taxonomy of 42 biophilic retrofit strategies for K-12 schools, as well as an aggregation of school case studies and health and performance literature reviews for the set of cards that were created within the taxonomy. The second stage was a user perception study to determine a strategy prioritization approach.

By analyzing both Kellert's 6 Biophilic Design Elements and Terrapin Bright Green's 14 Patterns of Biophilic Design, 42 specific biophilic design strategies were identified and included in the creation of a taxonomy based on their feasibility for implementation in an existing K-12 school setting. The strategies included in this set are uniquely developed, inspired by the previously published biophilic literature including Kellert and Terrapin Bright Green's frameworks. Factors for selection included the following: replicability in a K-12 school building given typical US conditions of infrastructure, space utilization, building codes and regulations; limiting the level of disruption that might be necessary for construction; and alignment with pedagogical goals of secondary education. First costs to implement and possible maintenance requirements, based on the literature findings and preliminary pricing, were also factors in the development of the taxonomy. In order to validate each strategy's impact on students and faculty, a rigorous literature review of potential health and performance impacts was conducted. Following the development of the biophilic retrofit taxonomy, a card set was created to engage school community stakeholders in the retrofit design of healthy and sustainable schools.

In order to evaluate the card set as a toolkit, a one-month long card sorting study was conducted between 4 March and 17 April 2022. This study investigated stakeholder perceptions of how beneficial and applicable the 42 different biophilic retrofit strategies were for the K-12 students and faculty. The community stakeholders included teachers, school board members, school administrators, parents, and school designers/consultants, who ranked the strategies based on perceived level of positive impact and affordability.

The ranking was achieved through a quadrant chart with the x-axis measuring affordability and the y-axis measuring level of positive impact (Figure 1). There were two different formats of the study, a paper version and an electronic version. Both were used in the overall analysis and conclusions of this study. This study was approved under Exempt Review by the Carnegie Mellon University Institutional Review Board (IRB) on 23 February 2022.

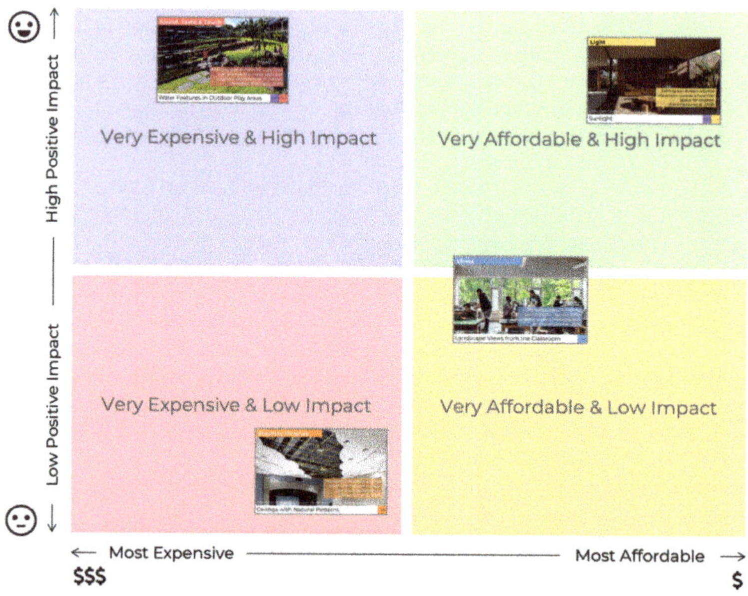

Figure 1. Quadrant chart for the card sorting game (Paper version) [15,38,39,45].

To ensure the accessibility of this study to the general public, a dedicated website was created to offer educational resources on biophilia and outline the focus of the research project. The website featured a link to the electronic survey and provided a downloadable version of the paper survey. Additionally, paper packages of the study were dispatched to four local K-12 schools in Pittsburgh. These packages included a comprehensive instruction packet, a set of the 42 cards, and a paper quadrant chart, which served as the metric for card ranking.

Identification of suitable local schools for the distribution of the paper version of the study was facilitated through a collaboration with the Green Building Alliance in Pittsburgh (GBA). Contacts recommended by the GBA received introductory emails outlining the study and its materials. Upon obtaining approval from the schools, paper packages were promptly mailed out. Participants in the paper-based card study were instructed to arrange the cards across the quadrant chart, placing those they deemed to have perceived positive impact for students and faculty onto the chart. Subsequently, participants were required to photograph their completed activity and submit it through the same survey link used for the electronic version. There was a separate section within the electronic survey for paper version submissions. The same survey was used to honor the IRB consent, and demographic questions were required for all participants. After completing the exercise and submission, the individual participants were asked to place the disassembled paper study back into the envelope and pass it forward to another K-12 stakeholder to extend the test set.

The second format consisted of an electronic survey. The survey software, Qualtrics, was used to create and administer the surveys. The electronic version of this study was aimed at reaching as many participants as possible, as well as collecting data from a variety of different regions and climates., The survey was shared to all relevant personal contacts of the author as well as advertised on a number of social media platforms including LinkedIn, Facebook, Instagram, and Reddit.

The electronic version of this study also asked participants to rank each card based on its level of impact as well as its affordability. Due to limited survey platform flexibility, participants were not asked to sort the cards on a quadrant chart via dragging and dropping; instead, they were asked to select the ranking of the cards on a scale of low, medium, and high for both impact and affordability (Figure 2). For the purposes of the card rankings, only the front of each card as well as the research summary that is listed on the back of the cards, was displayed. Participants could opt out of ranking a card if they felt there would be no level of positive impact.

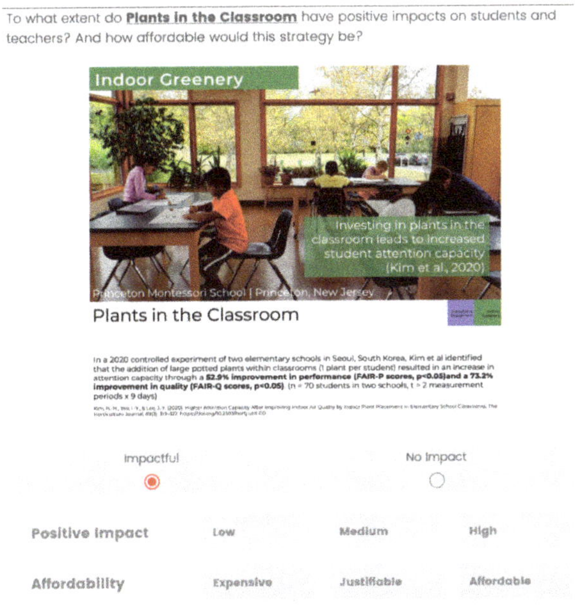

Figure 2. Electronic Study Survey Question on Plants in the Classroom [27].

Given these three levels of positive impact and affordability to choose from, a 3 × 3 matrix for card ranking was defined. Each box was assigned a score on a scale of −4 to +4. A score of +4 denotes a ranking of "High Impact and Affordable", while a score of −4 represents "Low Impact and Expensive". Additionally, it was decided that since affordability is very important to public schools, a strategy that is deemed to be "Medium Impact and Affordable" would score higher than a strategy that was "High Impact and Expensive". The same scoring applied to "Low Impact and Affordable" vs. "Medium Impact and Expensive" strategies (Figure 3).

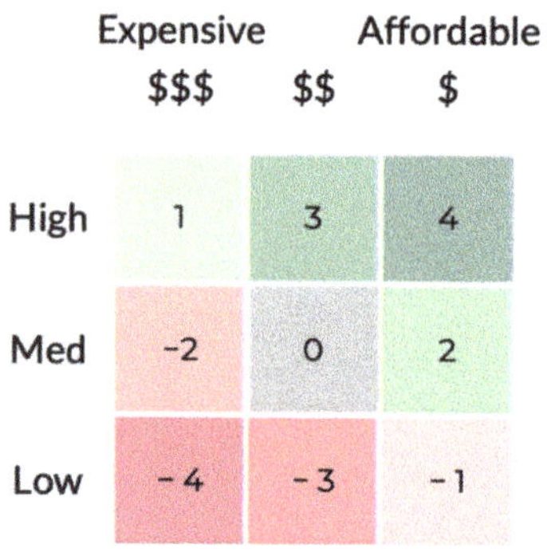

Figure 3. Scoring legend based on impact and cost for the card sorting study.

The average score for each biophilic action card is the value that was used to derive the statistical analysis in the results. Since the electronic version of the survey limited the ranking choices to a 3 × 3 matrix and the paper version was based on the quadrant chart, a 3 × 3 grid was manually overlayed on the images of the completed paper versions to make the paper and electronic results comparable in terms of the scoring.

The results of the card sorting exercise (both paper and electronic versions) were analyzed using Qualtrics, Microsoft Excel, R Studio, and Tableau. Descriptive statistics were utilized to understand key findings in the data. Statistical analysis was used to support the findings of the descriptive statistics using one-way ANOVA paired with Tukey's Procedure as a follow-up for unplanned corrections in the statistical analysis.

3. Results

3.1. A Taxonomy of Biophilic Retrofit Interventions in Cards

The taxonomy created in this study was strongly influenced by Terrapin Bright Green's 14 Patterns of Biophilia as well as Stephen Kellert's 6 Biophilic Design Elements. These strategies were identified to be both significant and feasible in a school setting through a comprehensive literature review. The final taxonomy consists of 42 biophilic retrofit strategies spread across seven categories (Figure 4). The seven categories are grouped into two parts, namely Visual Connection to Nature and Multisensory Connection to Nature. Within Visual Connection to Nature, there are three strategy sets as follows: Indoor Greenery, Views, and Biophilic Finishes. For Multisensory Connection to Nature, there are four factors, namely Natural Light, Nature's Sound, Taste, and Touch, Nature Interaction and Engagement, and Spatial Biophilic Strategies.

Based on this taxonomy, 42 cards were created with one card for each strategy (Figure 5). Each of the cards provided a brief summary of the strategy and a visual illustration of the strategy in practice on the front of the card, and the benefits of implementing the strategy based on a literature review, as well as a matrix of how the strategy aligned with the national standards and certifications on the back of the card (Figure 6). The standards and certifications included in the scope of the cards were LEED, WELL, CHPS (Collaborative for High Performing Schools), and LBC v4.0 (Living Building Challenge version 4.0).

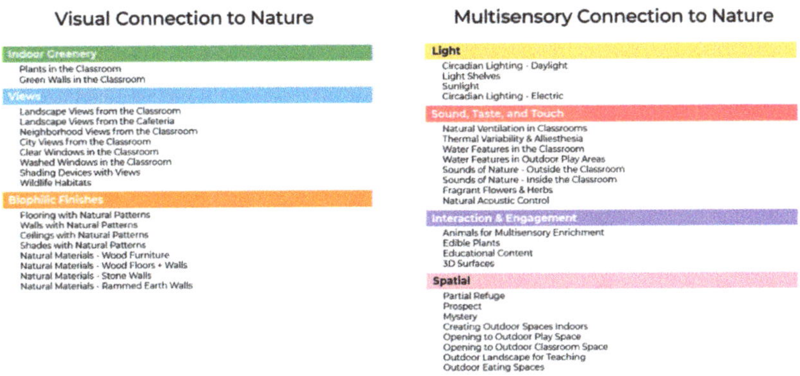

Figure 4. The taxonomy of the 42 biophilic retrofit interventions.

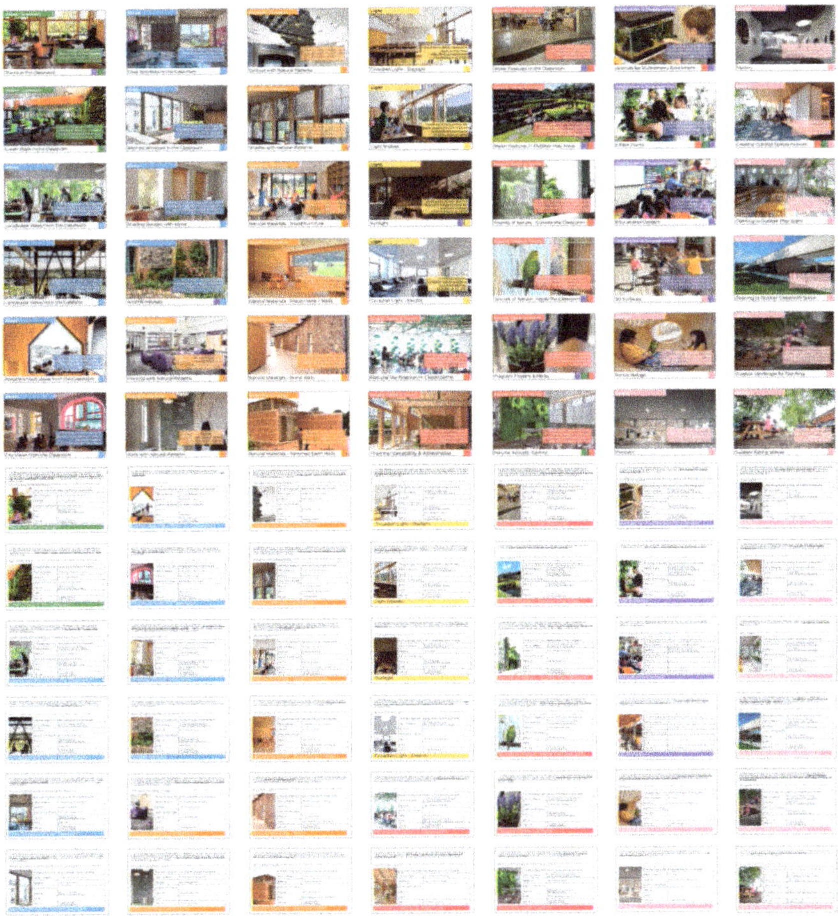

Figure 5. Full card set [12,13,15,18,19,24–47].

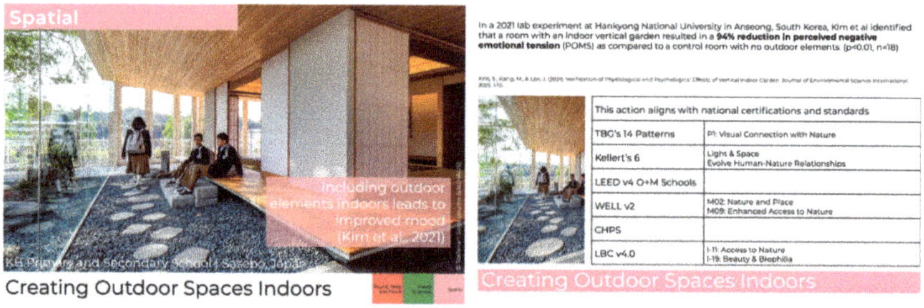

Figure 6. Creating Outdoor Spaces Indoors card front and back [20].

3.2. *Testing Stakeholder Perception of the Biophilic Taxonomy*

In order to evaluate the card set as a toolkit, a card sorting study was conducted in the spring of 2022. This study investigated the value of the 42 different biophilic retrofit strategies for K-12 students and faculty from the perspective of stakeholders, including teachers, school boards, school administrators, and parents, who ranked the strategies based on level of positive impact and affordability.

The card sorting study collected a total of 74 responses from the K-12 school community stakeholders. Of these 74 respondents, 78.4% were teachers and school administrators. Additionally, 54.8% of respondents were 45 years or older and 73.6% of all respondents were female. The online survey also collected responses from 13 of 50 states, which included at least one respondent from all the main geographical regions of the United States.

Each category was analyzed and the average score of each category was identified capturing the benefits of both impact and affordability (Table 1). The scores ranged from −4 to 4. Based on this, Interaction and Engagement, as a category, scored the highest with an average score of 1.67. As a category, Biophilic Finishes scored the lowest at −0.85. It should be noted that there were different numbers of cards in each category, with some ranked more impactful and affordable than others.

Table 1. Average biophilic category scores.

Category	Average Score
Interaction and Engagement	1.6695
Light	1.5154
Spatial	1.1231
Views	1.0307
Indoor Greenery	0.8565
Sound, Taste, and Touch	0.6613
Biophilic Finishes	−0.8528

While not every participant identified all 42 biophilic retrofit strategies in the card set as having any positive impact, all strategies had at least 77% of participants agree that there was some level of perceived positive impact on student and faculty outcomes. This is aligned with the current published research that appears on the back of the cards.

In addition to the overall category scoring analysis, subsets in each of the seven categories were separately analyzed to identify the strategies per category that were deemed the most impactful by the participants of this study. The cards within each category were compared to each other using statistical analysis based on the average scores of each card derived from the scoring approach which merges perceived level of impact and affordability as previously described (Figure 3).

In the Indoor Greenery category, plants in the classroom were identified to be the most impactful and affordable strategy for student and teacher outcomes at 95.9%, as compared to green walls at 37.5% ($p = 0.00$).

In the Views category, five of the eight strategies were identified to have a significant perceived impact on impact and affordability—clear glass windows, washed windows, landscape views from the classroom, wildlife habitats, and shading devices enabling views ($p = 0.0002$). Of these five, clear glass windows scored the highest with 75.3% of participants identifying it as having a high level of impact, and 98.6% deeming clear glass windows as having at least some level of positive impact on student and teacher outcomes.

Among the eight strategies in the Biophilic Finishes category, wooden furniture was identified to be the most impactful and affordable strategy for student and teacher outcomes ($p = 0.017$). In general, Biophilic Finishes had the most strategies being perceived by survey respondents as both low impact and expensive (−4). However, wooden furniture was perceived to be significantly more impactful and affordable than the other seven strategies with 47.8% of participants who identified the strategy as having a high impact level.

Of the four strategies in the Light category, both sunlight and circadian daylight were identified to have significant impact and affordability as compared to light shelves and electric circadian lighting ($p = 0.00$). Based on the results of the survey, 100% of all participants felt that sunlight would certainly have a positive impact on students and teachers, from low to high. This was the only strategy of the 42 strategies in the taxonomy that all 74 participants universally agreed would have positive impact.

Of the eight strategies in the Natural Sound, Taste, and Touch category, four were identified to have a significant positive impact and affordability—fragrant flowers and herbs, natural ventilation in the classroom, water feature outdoors, and sounds of nature from outside ($p = 0.00$). Of these four, fragrant flowers and herbs scored the highest with 55.7% of participants ranking the strategy as high impact.

Of the four strategies in the Interaction and Engagement category, both nature educational content and animals in the classroom were identified to have significant impact and affordability ($p = 0.00$). Of these two, nature educational content scored the highest with 48.5% of participants who ranked the strategy identifying it as a both high impact and affordable (+4).

Lastly, in the Spatial category, five of the eight strategies were identified to have significant impact and affordability—classroom opening to outdoors, opening to outdoor play space, outdoor landscape for teaching, partial refuge, and outdoor eating spaces ($p = 0.00$). Of these five, classrooms opening to the outdoors scored the highest with 83.3% of participants ranking the strategy as high impact, although only 12.5% felt it was both high impact and affordable.

In addition to the strategy ranking, the survey also asked stakeholders about additional biophilic retrofit strategies that were not included in the taxonomy and card set, that could be potentially added into the taxonomy. These strategies have yet to be reviewed and researched, but included:

- Natural patterns for air diffusers;
- Improving window area and view factors;
- Outdoor community gardens;
- Messy outdoor classrooms;
- Skylights;
- Dynamic light patterns (e.g., filtered/shaded skylights, programmable light systems);
- Material and tactile experience of nature.

The last analysis completed for this stakeholder perception study was an overall analysis on the entire taxonomy to identify which strategies US K-12 schools might prioritize based on the combined level of positive impact and affordability. A total of 74 K-12 school stakeholders identified eight strategies with the highest impact and affordability—Sunlight, Nature Educational Content, Circadian Daylight, Fragrant Flowers and Herbs, Outdoor Eating Spaces, Clear Glass Windows in the Classroom, Washed Windows in the Classroom, and Plants in the Classroom—as compared to 34 other biophilic retrofit design strategies ($p = 0.037$).

4. Discussion

In light of the severe underfunding of K-12 schools, decision makers must pinpoint cost-effective yet impactful design solutions for current school structures. Extensive academic research has demonstrated that biophilic design strategies offer both quantitative and qualitative advantages for human health and performance. This research project reveals that there exist numerous affordable retrofit strategies to introduce biophilic design into existing schools. It is crucial to involve the community stakeholders in K-12 schools in the decision-making process to ensure the creation of quality learning environments that foster the wellbeing and success of students and teachers.

Based on the results of the survey, these eight strategies could be prioritized for K-12 school retrofit planning and designs in order to affordably create more beautiful and impactful spaces for students and teachers—Sunlight, Nature Educational Content, Circadian Daylight, Fragrant Flowers and Herbs, Outdoor Eating Spaces, Clear Glass Windows in the Classroom, Washed Windows in the Classroom, and Plants in the Classroom. However, it should be noted that this prioritization assumes that windows are a given in an existing classroom, which is not always the case. Four of these eight strategies rely on access to a window in the classroom (Sunlight, Circadian Daylight, Clear Windows in the Classroom, and Washed Windows in the Classroom). This underscores the significance of windows in K-12 classrooms and provides additional justification for banning windowless classrooms in the future.

Within the full set of 42 biophilic retrofit strategies, there are certain strategies that may not be entirely compatible, such as pairing "refuge" or "mystery" with "prospect" (see cards in the Supplementary Material), or pairing shading devices with views, circadian daylight, or sunlight. However, the eight strategies deemed most impactful and affordable, as defined by the community stakeholders, could be incorporated together since they are not inherently conflicting.

The purpose of the 42 retrofit action cards was to help stakeholders make informed decisions about which strategies would be the most appropriate for their existing K-12 school. It was important for decision makers and K-12 school stakeholders to meet and discuss potential opportunities and limitations for each of the 42 strategies relative to their school.

This study was undertaken during the height of the COVID-19 pandemic and necessitated both physical and electronic engagement to increase the number of community stakeholders. Future studies would improve the online interface to enable nuanced location of the cards on the quadrant grid to match the paper version. The paper version of the survey facilitated greater dialogue and exchange, and a greater comparative shuffling of strategies—an altogether deeper level of engagement in the exploration of biophilic retrofits for their school.

5. Conclusions

The importance of high-quality school learning environments for the development and wellbeing of children cannot be overstated. Many K-12 schools, particularly public schools in the US, face significant funding challenges that impact the quality of their facilities. Given the substantial amount of time students spend within these environments, it is critical to address these deficiencies through strategic design and retrofitting efforts that enhance both student and teacher performance.

Biophilic design, which leverages the human affinity for nature, offers a compelling approach to improving school environments. This study has highlighted the substantial benefits of biophilic design in educational settings, supported by a robust body of literature that links a series of biophilic retrofit design strategies to improved cognitive function, reduced stress, better engagement, and overall enhanced academic performance.

Through the development of a taxonomy of 42 biophilic retrofit strategies and a subsequent user perception study, the most impactful and affordable strategies for K-12 schools were identified. These include sunlight, nature educational content, circadian daylight, fragrant flowers and herbs, outdoor eating spaces, clear glass windows, washed windows, and plants in the classroom. These strategies were consistently rated highly for impact and affordability by community stakeholders, indicating their feasibility and effectiveness in enhancing school environments.

It is essential for decision makers to prioritize these biophilic strategies in retrofit planning to create healthier, more productive learning spaces. While certain strategies may not be universally applicable due to existing infrastructure limitations, this taxonomy provides a flexible toolkit on biophilic retrofit strategies that can be adapted to the specific needs and constraints of individual schools. Engaging school community stakeholders in this process ensures that the selected strategies align with the unique educational and environmental goals of each institution.

Supplementary Materials: The following supporting information can be downloaded at: https://www.mdpi.com/article/10.3390/architecture4020024/s1, Figure S1: Printable card set.

Author Contributions: Conceptualization, K.L. and V.L.; methodology, K.L. and V.L.; formal analysis, K.L.; investigation, K.L.; writing—original draft preparation, K.L.; writing—review and editing, K.L. and V.L.; visualization, K.L.; supervision, V.L.; funding acquisition, K.L. All authors have read and agreed to the published version of the manuscript.

Funding: This research was funded by Carnegie Mellon University School of Architecture.

Institutional Review Board Statement: The study was conducted in accordance with the Declaration of Helsinki and approved by the Institutional Review Board of Carnegie Mellon University on 23 February 2022 (IRB ID: STUDY2022_00000031).

Informed Consent Statement: Informed consent was obtained from all subjects involved in the study.

Data Availability Statement: Data are unavailable due to privacy restrictions.

Conflicts of Interest: The authors declare no conflicts of interest. The funders had no role in the design of the study; in the collection, analyses, or interpretation of data; in the writing of the manuscript; or in the decision to publish the results.

References

1. Cochran, E. The Impact of School Building Occupancy, Facility Characteristics and Neighborhood Attributes on Student Academic Performance and Health Outcomes. Ph.D. Thesis, Carnegie Mellon University, Pittsburgh, PA, USA, 2014.
2. Filardo, M. *2021 State of Our Schools: America's PK–12 Public School Facilities 2021*; 21st Century School Fund: Washington, DC, USA, 2021.
3. American Association of School Administrators. School Budgets 101—AASA. Available online: https://www.aasa.org/uploadedFiles/Policy_and_Advocacy/files/SchoolBudgetBriefFINAL.pdf (accessed on 17 March 2022).
4. Alexander, D.; Lewis, L. *Condition of America's Public School Facilities: 2012–13 (NCES 2014-022)*; U.S. Department of Education. National Center for Education Statistics: Washington, DC, USA, 2014.
5. National Center for Education Statistics. Schools and Staffing Survey (SASS). Available online: https://nces.ed.gov/surveys/sass/tables/sass0708_035_s1s.asp (accessed on 1 April 2022).
6. Earthman, G.I. *School Facility Conditions and Student Academic Achievement*; UCLA's Institute for Democracy, Education, & Access (IDEA): Los Angeles, CA, USA, 2002.
7. ASCE. ASCE's 2021 Infrastructure Report Card. Available online: https://infrastructurereportcard.org/cat-item/schools/ (accessed on 1 April 2022).
8. Fromm, E. *The Anatomy of Human Destructiveness*; Macmillan: New York, NY, USA, 1964.
9. Wilson, E.O. *Biophilia*; Harvard University Press: Cambridge, MA, USA, 1984.
10. Kellert, S.R.; Heerwagen, J.; Mador, M. *Biophilic Design: The Theory, Science and Practice of Bringing Buildings to Life*; John Wiley & Sons: Hoboken, NJ, USA, 2008.
11. Browning, W.D.; Ryan, C.O.; Clancy, J.O. *14 Patterns of Biophilic Design*; Terrapin Bright Green LLC: New York, NY, USA, 2014.
12. Shen, J.; Zhang, X.; Lian, Z. Impact of wooden versus nonwooden interior designs on office workers' cognitive performance. *Percept. Mot. Ski.* **2020**, *127*, 36–51. [CrossRef] [PubMed]
13. Boubekri, M.; Lee, J.; MacNaughton, P.; Woo, M.; Schuyler, L.; Tinianov, B.; Satish, U. The impact of optimized daylight and views on the sleep duration and cognitive performance of office workers. *Int. J. Environ. Res. Public Health* **2020**, *17*, 3219. [CrossRef] [PubMed]
14. Ko, W.H.; Schiavon, S.; Zhang, H.; Graham, L.T.; Brager, G.; Mauss, I.; Lin, Y.W. The impact of a view from a window on thermal comfort, emotion, and cognitive performance. *Build. Environ.* **2020**, *175*, 106779. [CrossRef]

15. Fahimipour, A.K.; Hartmann, E.M.; Siemens, A.; Kline, J.; Levin, D.A.; Wilson, H.; Betancourt-Román, C.M.; Brown, G.Z.; Fretz, M.; Van Den Wymelenberg, K.; et al. Daylight exposure modulates bacterial communities associated with household dust. *Microbiome* **2018**, *6*, 175. [CrossRef] [PubMed]
16. Figueiro, M.G.; Kalsher, M.; Steverson, B.C.; Heerwagen, J.; Kampschroer, K.; Rea, M.S. Circadian-effective light and its impact on alertness in office workers. *Light. Res. Technol.* **2019**, *51*, 171–183. [CrossRef]
17. Loftness, V.; Aziz, A.; Choi, J.; Kampschroer, K.; Powell, K.; Atkinson, M.; Heerwagen, J. The value of post-occupancy evaluation for building occupants and facility managers. *Intell. Build. Int.* **2009**, *1*, 249–268. [CrossRef]
18. Gwak, J.; Shino, M.; Ueda, K.; Kamata, M. An investigation of the effects of changes in the indoor ambient temperature on arousal level, thermal comfort, and physiological indices. *Appl. Sci.* **2019**, *9*, 899. [CrossRef]
19. Zhang, Y.; Ou, D.; Kang, S. The effects of masking sound and signal-to-noise ratio on work performance in Chinese open-plan offices. *Appl. Acoust.* **2021**, *172*, 107657. [CrossRef]
20. Kim, S.; Kang, M.; Lee, J. Verification of Physiological and Psychological Effects of Vertical Indoor Garden. *Environ. Sci. Int.* **2021**, *30*, 1–10. [CrossRef]
21. Kellert, S.R. *Building for Life: Designing and Understanding the Human-Nature Connection*; Island Press: Washington, DC, USA, 2005.
22. Jarvis, I.; Davis, Z.; Sbihi, H.; Brauer, M.; Czekajlo, A.; Davies, H.W.; Gergel, S.E.; Guhn, M.; Jerrett, M.; van den Bosch, M.; et al. Assessing the association between lifetime exposure to greenspace and early childhood development and the mediation effects of air pollution and noise in Canada: A population-based birth cohort study. *Lancet Planet. Health* **2021**, *5*, e709–e717. [CrossRef] [PubMed]
23. Wijesooriya, N.; Brambilla, A. Bridging biophilic design and environmentally sustainable design: A critical review. *J. Clean. Prod.* **2021**, *283*, 124591. [CrossRef]
24. Chang, C.Y.; Chen, P.K. Human response to window views and indoor plants in the workplace. *HortScience* **2005**, *40*, 1354–1359. [CrossRef]
25. Fell, D.R. Wood in the Human Environment: Restorative Properties of Wood in the Built Indoor Environment. Ph.D. Thesis, University of British Columbia, Vancouver, BC, Canada, 2010.
26. Seo, J.Y. The effects of aromatherapy on stress and stress responses in adolescents. *J. Korean Acad. Nurs.* **2009**, *39*, 357–365. [CrossRef] [PubMed]
27. Kim, H.-H.; Yeo, I.-Y.; Lee, J.-Y. Higher Attention Capacity After Improving Indoor Air Quality by Indoor Plant Placement in Elementary School Classrooms. *Hortic. J.* **2020**, *89*, 319–327. [CrossRef]
28. Bernardo, F.; Loupa-Ramos, I.; Matos Silva, C.; Manso, M. The Restorative Effect of the Presence of Greenery on the Classroom in Children's Cognitive Performance. *Sustainability* **2021**, *13*, 3488. [CrossRef]
29. Gatersleben, B.; Andrews, M. When walking in nature is not restorative—The role of prospect and refuge. *Health Place* **2013**, *20*, 91–101. [CrossRef] [PubMed]
30. Marois, A.; Charbonneau, B.; Szolosi, A.M.; Watson, J.M. The Differential Impact of Mystery in Nature on Attention: An Oculometric Study. *Front. Psychol.* **2021**, *12*, 759616. [CrossRef] [PubMed]
31. Roskams, M.; Haynes, B. A randomised field experiment to test the restorative properties of purpose-built biophilic "regeneration pods". *Corp. Real Estate* **2020**, *22*, 297–312. [CrossRef]
32. Kuo, M.; Browning, M.H.; Penner, M.L. Do lessons in nature boost subsequent classroom engagement? Refueling students in flight. *Front. Psychol.* **2018**, *8*, 2253. [CrossRef] [PubMed]
33. Matsuoka, R.H. Student performance and high school landscapes: Examining the links. *Landsc. Urban Plan.* **2010**, *97*, 273–282. [CrossRef]
34. Africa, J.; Heerwagen, J.; Loftness, V.; Ryan Balagtas, C. Biophilic design and climate change: Performance parameters for health. *Front. Built Environ.* **2019**, *5*, 28. [CrossRef]
35. Ansari, A.; Pettit, K.; Gershoff, E. Combating obesity in head start: Outdoor play and change in children's BMI. *J. Dev. Behav. Pediatr. JDBP* **2015**, *36*, 605. [CrossRef] [PubMed]
36. Romar, J.E.; Enqvist, I.; Kulmala, J.; Kallio, J.; Tammelin, T. Physical activity and sedentary behaviour during outdoor learning and traditional indoor school days among Finnish primary school students. *J. Adventure Educ. Outdoor Learn.* **2019**, *19*, 28–42. [CrossRef]
37. Osama Mohammed Fikry, N.; ElFiki, S. Teachers' Interpretation of "Learning through Landscapes" in Egyptian School Grounds. *Engineering Research Journal* **2020**, *168*, 325–342. [CrossRef]
38. Kuhlenengel, M.; Waters, C.E.; Konstantzos, I. Assessing the impact of outside view on learning: A close look to EN 17037 'view out' practices through the analysis of 220 classrooms. *J. Phys. Conf. Ser.* **2019**, *1343*, 012159. [CrossRef]
39. Determan, J.; Akers, M.A.; Albright, T.; Browning, B.; Martin-Dunlop, C.; Archibald, P.; Caruolo, V. *The Impact of Biophilic Learning Spaces on Student Success*; American Institute of Architecture, Building Research Knowledgebase: New York, NY, USA, 2019.
40. Heschong, L.; Wright, R.L.; Okura, S. Daylighting impacts on human performance in school. *J. Illum. Eng. Soc.* **2002**, *31*, 101–114. [CrossRef]
41. Green Bronx Machine. Available online: https://greenbronxmachine.org/about/ (accessed on 8 November 2021).
42. Junge, R.; Wilhelm, S.; Hofstetter, U. Aquaponic in classrooms as a tool to promote system thinking. In Proceedings of the Conference on Agriculture, Environmentalism, Horticulture, Floristics, Food Production and Processing, Strahinj, Naklo, Slovenia, 14–15 November 2014.

43. Damián-Chávez, M.M.; Ledesma-Coronado, P.E.; Drexel-Romo, M.; Ibarra-Zárate, D.I.; Alonso-Valerdi, L.M. Environmental noise at library learning commons affects student performance and electrophysiological functioning. *Physiol. Behav.* **2021**, *241*, 113563. [CrossRef] [PubMed]
44. Brambilla, A.; Jusselme, T. Preventing overheating in offices through thermal inertial properties of compressed earth bricks: A study on a real scale prototype. *Energy Build.* **2017**, *156*, 281–292. [CrossRef]
45. Herrington, S.; Brussoni, M. Beyond physical activity: The importance of play and nature-based play spaces for children's health and development. *Curr. Obes. Rep.* **2015**, *4*, 477–483. [CrossRef] [PubMed]
46. Newman, O. *Creating Defensible Space*; Diane Publishing: Collingdale, PA, USA, 1996.
47. Drown, K.K.C. *Dramatic Play Affordances of Natural and Manufactured Outdoor Settings for Preschool-Aged Children*; Utah State University: Logan, UT, USA, 2014.

architecture

MDPI

Article

Ascending to Well-Being through Mediated Spaces: An Alternative to Informal Learning and Physical Activity Environments in Vertical Schools

Elia Ebrahimi Salari * and Nigel Westbrook

UWA School of Design, The University of Western Australia, Crawley, WA 6009, Australia; nigel.westbrook@uwa.edu.au
* Correspondence: elia.ebrahimisalari@research.uwa.edu.au

Abstract: The correlation between health and well-being, outdoor activity, and the natural environment in learning environments has been recognised by pioneers like Samuel Wilderspin and modern theorists like Maria Montessori, who have underscored the importance of integrating nature into school designs to foster connections between students, teachers, and the community, thereby promoting physical, social, and mental well-being. However, for schools in hot–humid climatic regions in densely populated cities of Southeast Asia, reduced time spent in nature due to adverse climatic conditions has led to an increasingly inactive lifestyle, impacting children's health and well-being. Overpopulation in major cities, such as those in Southeast Asia, further exacerbates this challenge, leading to a scarcity of land and the rise of high-rise buildings, including vertical schools. This article explores the evolving learning environment for children, emphasising the critical correlation between well-being, nature, and school settings. It evaluates case studies of best practices in learning environment design, focusing on how architecture can support pedagogical goals. The analysis identifies spaces that affect well-being, termed mediated spaces, and recommends guidelines for such spaces tailored to vertical schools in hot–humid climates. The methodology includes a literature review of learning theories, the integration of learning environments with nature, and biophilic design. This review forms the basis for developing adaptable design guidelines tailored to hot–humid climates. Additionally, case study analyses of exemplary schools are conducted to identify mediated spaces that enhance well-being and adapt these findings to vertical school designs.

Keywords: learning environment; vertical school; mediated spaces; natural exposure; biophilic; health and well-being; social interactions; outdoor activities; pedagogy; informal learning

1. Introduction

The idea of integrating schools with nature and community is not a new phenomenon. There is no doubt that the discovery of childhood dates back to the 13th century, and its development can be tracked in the history of art in the 15th and 16th centuries. However, from the end of the 16th century and throughout the 17th century, its development became more evident [1] (p. 47). By the late 18th century, the concept of the child was firmly established, and childhood discovery space was formalised as an "institution" [1] (p. 334). This institution was seen as a "walled garden" in which small and weak children were protected from "the harshness of the world outside until they became strong and clever enough to cope with it" [2] (pp. 8–9). In the early 19th century, the educational garden concept commenced with kindergartens established by Friedrich Froebel, the German educator who invented the kindergarten, emphasising the importance of integrating children's spaces with the natural environment [3] (p. 200).

Nevertheless, nowadays, children spend less time in natural settings than was normal in the past as a consequence of urban competition for space, which is creating more high-rise buildings, including new typological models for schools [4] (p. 4) [5]. Traditional school typologies, which have been constructed in a horizontal, scattered manner, are intended to provide a space for learning and development and a balance between mental and physical developmental activities. However, as the city's population expands, the land becomes scarce, and developing a comfortable learning environment becomes more challenging as schools become vertically planned. A growing population, urban sprawl, densification strategies, and digital integration are indicators of time spent in and out of natural environments, which undeniably impact children's health and well-being. Given the close correlation between time spent in nature and better community well-being, it is of concern that people spend more than 80% of their time in buildings deprived of nature's opportunities. As a result, one of the necessary elements for encouraging progressive thinking in educating children is to create a functioning and supportive physical learning infrastructure in vertical buildings [5,6] (p. 13).

Furthermore, bringing the outside in and naturalising play spaces could lead to connection to the greater community and landscape beyond the school gate, providing unlimited exploration and discovery opportunities. Moreover, having less activity has also been cited as one of the reasons for child obesity and obesity-related diseases like diabetes in the future. Obesity in children rose to about 20% in 2004 from about 4% in the 1960s [7] (pp. 75, 78, 97).

The lack of socialisation has been argued as another source of mental health effects. According to a document by the government of Australia, the risk of mental illness can be reduced by having people to talk to, rely on, and make new connections through hobbies or social groups [8]. In addition, the importance of social interactions could be seen in Dewey's philosophy of "schools as social institutions" and "education as a process of living". However, nowadays, many of our schools are disconnected from their communities due to commuting time on long highways, limited hours, and inflexible spaces, and they miss out on becoming part of a social ecosystem that can enrich both students and communities [7] (pp. 108–109).

According to World Bank statistics, of the 7.4 billion people making up the world's population, 4 billion live in urban areas, a numerical figure that is expected to increase significantly to 6 billion by 2045. Accommodating this massive number of people requires, in turn, more innovative design strategies for cities [9] (p. 3).

The emergence of vertical schools of four or more storeys in height in Australia's capital cities dates back to 1976 with the eight-storey St Andrew's Cathedral School in Sydney [10] (p. 86). While "vertical schools are a relatively recent phenomenon in Australia, developed in response to changing demographics within the centres of our capital cities" [11] (p. 29), the construction of vertical schools is more common and has a long history in Europe, Asia, and America [10] (p. 86).

Swinburn argues that, although vertical schools that are usually between 4 to 17 storeys high are a new trend in densely populated cities, there is still a lack of adequate policies and regulations to propose a set of appropriate design strategies to apply to them [10] (p. 86) [12] (p. 20). Similarly, Bogle Architects noted in relation to their recent vertical school project in Singapore in 2017, the Early Learning Village, that there was no "rule book" for constructing a project of this typology. As a result, it can be seen that the majority of such projects have relied on fundamental principles, guesswork or intuition, or the study of precedents, analysing the advantages and disadvantages of previous attempts to design a vertical school [13].

The above gaps in knowledge are even more challenging when it comes to the application of vertical building within densely populated cities of Southeast Asian countries in hot–humid climates. In such hot–humid climates, the impact of climatic limitations may result in children's activity being restricted because of the lack of activity spaces in a vertical building. In recent decades, for schools in hot–humid climatic regions, reduced time spent in nature and outdoor activities due to the pathogenic effects of exposure to such climatic conditions and, as a consequence, an increasingly inactive lifestyle are factors that have posed significant consequences for children's physical, social, and cognitive health and well-being [4] (p. 4).

Following the above statements, this research proposes a set of design guidelines for designing "active spaces" in primary-level vertical schools in hot–humid climatic regions. These active spaces are termed "mediated spaces" in this research. They include semi-open spaces like sheltered rooftops, patios, internal courtyards, and terraces, as well as transitional spaces such as corridors, foyers, lobbies, atriums, and staircases [14] (p. 150). Mediated spaces as interaction spaces to integrate indoor and outdoor environments in vertical schools can improve adjacent spaces' experiential quality, create an interactive and physical environment, increase people's perception of comfort, and, consequently, improve children's well-being. These mediated spaces may be understood as what David Leatherbarrow has termed "dwelling" equipment - spaces that interact with human subjects to engender a sense of place formation. They constitute a significant differentiation of vertical schools from typical horizontal schools by creating the required natural, interactive, and physical spaces in a vertical arrangement [15] (p. 119). So, the main question that this article addresses is as follows: how can spaces mediate interaction between natural spaces and children's physical activity?

2. Learning and Pedagogy

2.1. What Is Learning?

Learning has been defined as "a process that leads to change, which occurs as a result of experience, and increases the potential for improved performance and future learning" [16] (p. 3). This definition underscores several crucial aspects of learning. Firstly, it highlights that learning is a process rather than a fixed outcome. Secondly, it stresses that learning entails changes in knowledge, beliefs, behaviours, and attitudes. Lastly, it emphasises that learning is not imposed upon students but is rather a direct outcome of how they perceive and react to their experiences. It is a direct consequence of how students interpret and respond to their experiences [16] (p. 3).

Raffi Cavoukian, a Canadian troubadour and the founder of "Child Honouring", an integrated philosophy linking person, culture, and planet, argues that children can learn from anything they are immersed in, not just from what they are being taught at that moment [7] (p. 30). However, without basic needs being met, learning cannot begin. According to Abraham Maslow, one of the founders of humanistic psychology, "people were guided by their needs, and as soon as one need was satisfied, they would move on to the next" [7] (p. 34). In his "Hierarchy of Needs" ladder, he outlines five stages of human development, which are shown in Figure 1. These stages begin with primary physiological and safety needs, followed by belonging/love and self-esteem, and end with self-actualisation. A lack of these requirements can hamper a child's performance at home, school, and in adulthood [7] (pp. 34–35).

Figure 1. Maslow's Hierarchy of Needs ladder. Source: [7] (p. 34).

2.2. Children's Developmental Aspects

In this section, several theoretical perspectives of seminal educationalists, including Dewey, Montessori, Piaget, Malaguzzi, and Vygotsky, that emphasise different aspects of a child's development are discussed.

Dewey's perspective that "children learn by experience or learning by doing" emphasises the physical development of children. In her last book, *The Absorbent Mind*, Maria Montessori similarly argues that "mental development must be connected with movement and be dependent on it. It is vital that educational theory and practice should be informed by that idea" [17] (p. 26).

Piaget, in his "cognitive development theory", classifies these developmental stages of a child's brain according to age. Children begin to think in an increasingly complex pattern between the ages of two and seven. They begin to understand concepts such as symbols and time and then understand their individual emotions at the age of seven to eleven when they reach the concrete operational stage [17] (p. 17). In addition, Montessori argues that "cognitive development involves kinaesthetic movement, both fine motor and gross motor" [17] (p. 28). These developing kinaesthetic skills are influenced and assisted by appropriate learning environments. Malaguzzi argues, for example, that for cognitive development, "by designing the spaces that create a pleasant learning environment for the child, space itself can become a third teacher for the students" [17] (p. 30).

Human interactions are fundamental to social development as the child experiences them over time [17] (p. 23). In Vygotsky's child development theory, the fundamental role of social interaction in the development of cognition is stressed [18] (p. 131). Dewey also talks about the social dimension of learning: "I believe that school is primarily a social education. I believe that education, therefore, is a process of living and not a preparation for future living" [7] (p. 108). The emphasis on the social dimension of learning has implications for the physical learning environment [19] (p. 117).

According to the mentioned theoretical perspectives, children's development could be categorised into three main aspects, namely, physical development (playing and exercising), cognitive development (learning), and social development (forming and sustaining relationships), as illustrated in Figure 2.

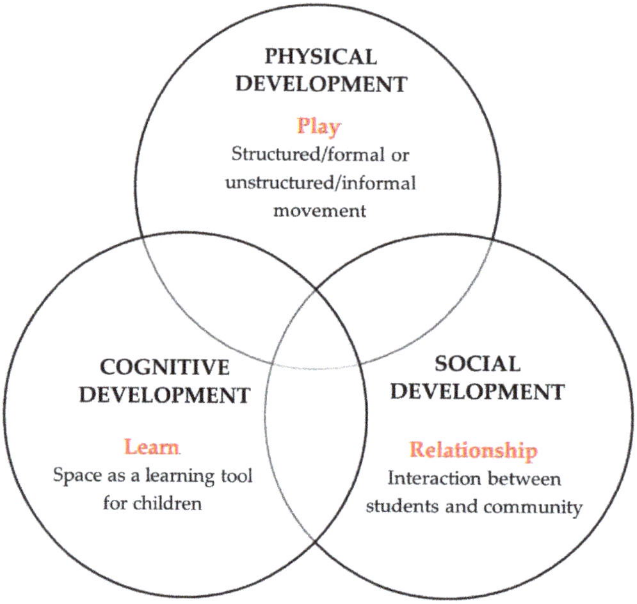

Figure 2. Different aspects of a child's development. Source: author.

2.3. From Traditional to Innovative Pedagogy

In pedagogical environments, there are two main approaches to learning. Traditional learning relies on teacher-centred methods such as lectures, with students primarily listening and taking notes. In contrast, active learning relies on student-centred techniques to engage students and make them active participants in their education. This often involves teamwork in cooperative learning settings. Cooperative learning is also a subset of active learning, emphasising collaborative tasks within small groups of two to five students, where each student is individually responsible for both their own and the group's learning. While traditional lecture formats are criticised for promoting student passivity and struggling to keep up with the evolving information landscape, active and cooperative learning are associated with various benefits, including improved student attitudes, higher academic achievement, enhanced comprehension, retention, transference of learning, and the development of advanced thinking skills [20] (pp. 10–11).

School buildings have been categorised according to different aspects and aims of the design. In one example, Kim Dovey and Kenn Fisher demonstrated a definition of spaces required in modern school design (based on constructivist pedagogies) in a systematic review of 59 award-winning international school designs in 2014. As can be seen from Table 1, Dovey and Fisher have classified the different spatial types found within the learning clusters into six primary categories. A traditional classroom is the starting point but is further extended to what is called the "commons", "streetspace", "meeting", "fixed function", and "outdoor learning". However, meeting areas and fixed functions can be integrated into commons, streetspaces, and classrooms or separated [21] (pp. 43, 46, 48).

Finally, they grouped their findings into five genres of learning space design, which they described as a "five-part typology" [22] (p. 11). Even though not all plans can be neatly classified into the five types of clusters above, Dovey and Fisher suggest that this can be a helpful framework for analysing the range of spatial experiments that are taking place to meet changing pedagogy needs [21] (pp. 52–54). Figure 3 shows the adapted content of Dovey and Fisher in an ILETC survey by Professor Wesley Imms [22] (p. 12).

Table 1. Typology of learning spaces by Dovey and Fisher. Source: [21] (p. 48).

Spaces	Definitions
Classrooms	A traditional closed learning space of about 40–60 sqm for 20–30 students. If learning spaces are fully closable with flexible walls, then they are classified as classrooms.
Commons	A learning space greater than about 40 sqm that cannot be fully closed into 25 student classrooms (or smaller) and is not the major access route to any other commons or classroom, hence protecting it from major through traffic.
Streetspace	An open learning space about 3 m in width (allowing activity + circulation) that cannot be closed into classrooms and is exposed to major through traffic as the primary access space to other learning spaces.
Meeting area	A small learning area of less than 40 sqm accommodating groups of 5 to 20. While such spaces may house seminars, the key criterion is that they cannot house a traditional class size.
Fixed function	Any learning space fitted for specialised use such as "Arts", "Science", "IT", "Computers", "Wet Area", "Music", "Drama", and "Resources".
Outdoor learning	Any outdoor area defined on the plan as an integral part of the learning cluster, generally labelled "outdoor learning", "outdoor room", or "learning court". Simple access to the outdoors does not qualify.

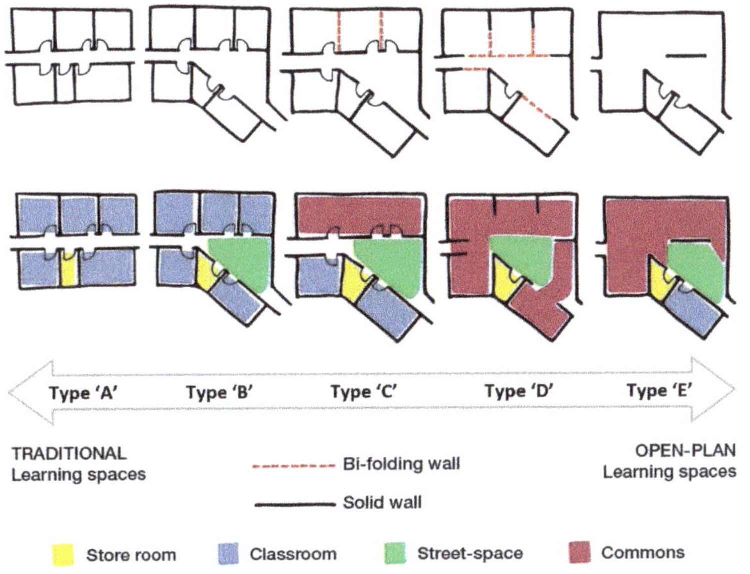

Figure 3. Dovey and Fisher's learning spaces typologies. Source: [22] (p. 12).

These learning space typologies demonstrate how different pedagogies and educational methods affect architecture and space and range from the traditional classroom to various degrees of convertibility to completely open plans [22] (pp. 52–54):

- Type A: Traditional closed classrooms entered by a corridor or access space without direct access to other teaching spaces and openability between classrooms.
- Type B: Traditional classrooms with breakout space. Streetspace has been introduced into the teaching/learning cluster, while the classrooms have remained intact without any commons.
- Type C: Traditional classrooms with flexible walls and breakout space. "Convertible classrooms" are introduced here as learning clusters where flexible walls allow two or more traditional classrooms (and possibly adjacent meeting rooms and wet areas) to be combined into a single common. By using such an approach, a wide range of

pedagogies can be employed while maintaining the reversibility of the traditional classroom.
- Type D: Open plan with the ability for separate classrooms. This type of "convertible streetspace" is used as it includes plans that open classroom clusters to the street and each other, creating a more extensive common.
- Type E: Open plan with some adjoining spaces. This type is characterised by "dedicated commons", where a protected "commons" serves as the spatial core of a learning cluster because it cannot be transformed into closed classrooms without extensive renovations [22] (pp. 52–54).

These typologies trigger the concept of spaces having more than one function, thus making them flexible, and groups of one genre being nested within another, which will be discussed later at the end of this article.

3. Learning Environment and Nature

3.1. Children's Problems in Today's Learning Environment

There is substantial evidence that outdoor play promotes physical, social, and mental well-being. Organ growth and muscle building are among the physical well-being benefits of outdoor play. Exploring children's local neighbourhoods and learning the rules of everyday life are the social well-being benefits of outdoor play. Improving children's emotional and academic development are the mental well-being advantages of outdoor play [23] (p. 6).

However, as seen in the last few decades, reduced time spent in nature and outdoor activities, along with an increasingly inactive lifestyle and indoor, sedentary, recreational activities, may have significant consequences for children's physical, social, and cognitive health and well-being. As a result, inactive lifestyles can influence physical health, increasing the risk of obesity and affecting children's cognitive performance and relationships [4] (p. 4) [24] (pp. 24, 26).

Richard Louv, the author of *Last Child in the Woods*, coined the term "nature-deficit disorder" to describe the lack of free-ranging children's discovery of "wildlands" in towns. He also explained possible adverse effects on human well-being and social cohesion as children move indoors and away from direct interaction with the natural environment [25] (p. 433). Indeed, according to the book *School Facilities: America's Schools Not Designed or Equipped for the 21st Century*, educators and students spend a great deal of time indoors (85% to 90% mostly at home and at school), while indoor pollution, such as cleaning supplies, air fresheners, and personal care products, is usually much higher than the amount of pollution outdoors, sometimes by as much as ten or even a hundred times [7] (p. 27).

In addition, it should be noted that nowadays, vertical schools present specific challenges associated with children's connection to the outdoors, and due to living in an urban environment, these connections are severely restricted for the students [17] (p. 4). So, to avoid all the above issues of children in today's vertical schools, more outdoor interaction opportunities should be considered. This study seeks to propose typologies for ways of providing nature interaction in high-rise schools.

3.2. Background of Integrating Education, Nature, and Community

The beneficial effects of the interaction of students with the natural environment have been long recognised. Samuel Wilderspin (1792–1866), an English educator, and David Stow (1793–1854), a Scottish educator, both established education systems supported through school building laws and recognised the need for an outdoor playground area [26] (pp. 539–540). The context at this time was, of course, the industrial revolution, water-borne diseases caused by overcrowding, and uncontrolled atmospheric pollution. In the twentieth century, John Dewey also introduced an evolutionary philosophy based on the idea that "children learn by experience" [17] (p. 26). He argued that "the school is primarily a social institution" and that "education... is a process of living and not a preparation for future living". To achieve this, he argued that interaction with open-space environments should be considered a priority in designing learning environments for children. The eminent Italian educational theorist, Maria Montessori, advocated that mental growth must be related to and based upon movement. She also noted an almost mathematical correlation between the child's environment, activity, and development [7] (pp. 80–81, 97, 108).

The importance of educational theory cannot be overstated, as it also impacts design theories. For example, the Austrian–American modernist architect Richard Neutra, an advocate for the health benefits of exposure to the sun and fresh air, argued that schools must interact with the existing site. In his buildings, he attempted to integrate both the visual and physical elements of nature [27] (p. 21). Similarly, Loris Malaguzzi's Reggio Emilia concept proposed that "by designing the spaces that create a pleasant learning environment for the child, space itself can become a third teacher for the students". Malaguzzi contended that "space becomes a learning tool for the children [17] (p. 30).

According to David Suzuki, an award-winning scientist, environmentalist, and broadcaster, one of the most critical things today's children need is "reconnection with nature", as we live in a world that is mainly made by humans with a few plants and pets around us. Plants, trees, flowers, water, animals, and insects together form elements of children's perceptions of nature [7] (pp. 96, 140). Louv also believes that the lack of contact with nature in childhood diminishes senses, and such disassociation from nature exacerbates attention difficulties [24] (p. 25). As schools are the last opportunity for children to interact with nature, the term "schoolyard greening" has emerged to explain how natural habitat restoration is being implemented in school environments. Planting trees and creating vegetable gardens are some examples of bringing nature back to the school. As part of nature-based learning, children also get the chance to learn about nature and discover who they are as a part of nature [24] (p. 26).

3.3. *Vertical School as a Learning Environment Type*

All the above school examples were less than four levels, mainly being one or two-level buildings, and all mediated between classrooms and larger contexts, either the natural setting or more socially interactive spaces, like Louis Kahn's ideas of buildings as communities and of nesting cells (of classrooms, for example) within larger collective units. So, the question is this: how can you create such mediations and nesting within the context of a vertical school?

One possibility that comes to mind is based upon the retail precept that customers do not want to change levels more than once. So, you could have three levels clustering together, but the vertical circulation system connects up all the clusters. Each cluster could perhaps accommodate a couple of year groups, consisting of a developmental stage. By the end of this article, this concept will be expanded; meanwhile, let us start by defining what differentiates vertical school from horizontal school and how it emerged.

Regarding the management of movement and circulation/building morphology and form in an urban environment, which implies the height of a building, schools can also be categorised into horizontal and vertical typologies. Both vertical and horizontal schools have their merits and are used in various educational settings worldwide; however, vertical schools are often built to meet the needs of urbanisation, densification, and sustainable development [28] (pp. 6, 11–12).

Schools are typically built in suburban areas, spread over long distances, and one or two storeys in height, with children usually walking long distances between wings or around courtyards. The land is used extensively in these settings, and the ecosystem is significantly impacted [29] (p. 40). Significantly, the idea of the school as a demonstration of sustainability has only recently been attempted. In the past, the landscape was used for sports, privacy, or a minimal amount of beautification—that is to say, in a functional or decorative manner but rarely as an ecological demonstration.

As urban densification is inextricably connected to today's lifestyles, a more widespread engagement with the idea of vertical school (VS) is now required in many urban areas [30] (p. 181). A vertical school typically occupies urban infill areas, has a small footprint, and relies on stairs and lifts for circulation. The smaller floor plates allow access to light and the exterior, while visual connections between floors encourage interaction between occupants [29] (p. 40). A vertical school is designed as a learning environment for all teaching, management, and leisure activities. It is contained in one or two buildings, is 4 to as many as 17 floors high, with elevated outdoor areas replacing the traditional greenfield surrounds [10] (p. 86). A vertical school saves space and takes up less land, so more land can be used for developing other ventures. It enables students to be closer to the resources they need for their future careers, as they can be located in high-population centres. Proximity to the urban core leads to more intimate relationships with local businesses, creates more opportunities, and occupies less space, thereby helping with overcrowding by increasing the density of infrastructure [6] (pp. 14, 16–17). It can also allow the school to contribute amenities to the local district through greater use of its resources (landscape, halls, library, etc.). Multi-storey schools, however, require innovative architectural and pedagogical approaches to ensure direct access to nature and green areas. The prevalent problems in every vertical school are the lack of natural playgrounds and schoolyards, outdoor workouts, and physical activity spaces [30] (pp. 181, 183). Figure 4 illustrates the various layers and contexts in which learning occurs, encompassing a diverse range of environments. Vertical schools are therefore considered a subsection of this broader layer.

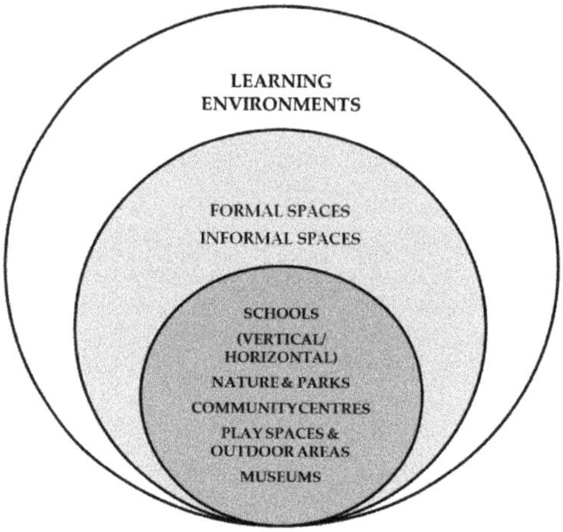

Figure 4. Different layers of learning environments. Source: author.

While vertical schools have a well-established international history and are becoming more prevalent locally, there is a notable scarcity of evidence-based research concerning these institutions. The lack of scholarly literature has led to a significant volume of discussions and opinions about vertical schools. Some of this commentary is critical of vertical schools, expressing concerns about students' well-being due to limited access to nature and physical activity opportunities. Conversely, other discussions praise these schools for their potential to spearhead educational reform or view them as a vital response to urban densification [28] (p. 6).

3.4. Biophilic Design

The concept of resilience has emerged as an essential urban aspiration and as a parallel concept to sustainability (sometimes replacing it). Sustainability has become an increasingly important goal and frame of reference for cities over the last two decades. Sustainability is understood as an integrated framework for guiding city development and helping cities accomplish many tasks at the same time, such as reducing ecological footprints and resource demands, deepening connections to place and landscape, and enhancing liveability and quality of life while expanding economic opportunities for the least privileged, among other things. Resilience is a term with many meanings, but at its core, it refers to the ability to adapt and respond to shocks in a positive manner; it comes from the Latin "resiliere", which means to jump back or rebound. Godschalk defines a resilient city as one that can overcome severe shocks without suffering immediate chaos or permanent damage. In order to be resilient, a community must adapt to dynamic social and ecological conditions in a way that optimises quality of life, long-term ecological productivity, and public and personal health as a whole [31] (pp. 3331–3332).

In order to maintain resilient design and health and well-being in schools, one approach is considered in this section—the use of biophilic design. The term "biophilia" was used for the first time by Edward Wilson in his article "Biophilia" in 1979 [32] (p. 43). He described biophilia as "the innate tendency to focus on life and lifelike processes" [32] (p. 1). Newman defined biophilia literally as "love for our living systems" [33] (p. 47). According to the concept of biophilic design, excellent design must embrace nature and natural components at the building, site, city, and regional levels [31] (p. 3328). Biophilic design involves building with natural materials, natural light, vegetation, natural views, and other natural elements. This can improve indoor air quality and positively impact students' learning, including indoor planting and green walls [34] (pp. 8, 22).

Beatley and Newman have argued, on the basis of evidence, that biophilic cities or biophilic urbanism will enhance or increase urban resilience: "The movement in the direction of making cities greener, more natural, more biophilic, will also help to make them more resilient" [31] (pp. 3331, 3333). Newman describes the benefits of biophilic design in cities. In addition to cooling the city (especially with the increase in urban heat island and climate change), the rain is slowed down as it does in a forest, which reduces stormwater surges. Other benefits include reducing energy consumption in buildings due to the mantle of insulation from plant life and improving biodiversity and health [33] (p. 47).

Table 2 summarises the elements of biophilic city designs at different scales by Beatley and Newman [31] (p. 3330). Building, block, and street scales are the focus areas of this research study.

Table 2. Beatley and Newman's elements of biophilic city design at various scales. Source: [31] (p. 3330).

Scales	Biophilic Design Elements
Building	Green rooftops Sky gardens and green atria Rooftop garden Green walls Daylit interior spaces
Block	Green courtyards Clustered housing around green areas Native species yards and spaces
Street	Green streets Urban trees Low impact development (LID) Vegetated swales and skinny streets Edible landscaping High degree of permeability
Neighbourhood	Stream daylighting, stream restoration Urban forests Ecology parks Community gardens Neighbourhood parks/pocket parks Greening grey fields and brownfields

Table 2. *Cont.*

Scales	Biophilic Design Elements
Community	Urban creeks and riparian areas Urban ecological networks Green schools City tree canopy Community forest/community orchards Greening utility corridors
Region	River systems/floodplains Riparian systems Regional greenspace systems Greening major transport corridors

Biophilic design objectives focus on incorporating nature and natural elements into the built environment with the objective of enhancing well-being and the connection of occupants with nature. As biophilic objectives are related to health and well-being in built environments, they also pertain to the earlier discussion in this section.

4. Definition of Mediated Spaces

This section focuses on finding different types of mediated spaces and their application in learning spaces. Physical activities and interaction with the natural environment in vertical schools with limited land can occur in mediated or in-between spaces. Mediated spaces are defined as spaces where interaction between indoor and outdoor environments happens [35] (p. 32). In terms of pedagogy, mediated spaces can be defined as spaces where we can shift from formal learning to informal learning. This informal learning can take place in settings including semi-open spaces like patios, internal courtyards, semi-closed rooftops, and terraces, as well as transition spaces like corridors, foyers, lobbies, atria, and staircases. These mediated spaces facilitate knowledge sharing and peer-to-peer learning in the form of "chance encounters and social interactions". Moreover, learning in informal settings like corridors, breezeways, and circulation zones and expanding mediated spaces outdoors can enhance learning ability and well-being and build relationships in education settings [5].

Transitional spaces, as a subcategory of mediated spaces, are characterised as both buffer spaces and physical relations between outdoor and indoor environments. The design of these spaces is considered very important by architects for aesthetic reasons and emergency purposes, such as providing escape routes in the event of a fire, apart from being practical as circulatory routes into and throughout the building. These would have advantages for inhabitants going in and out of places, reducing thermal shocks and shifting comfort expectations. On the other hand, transitional spaces will typically require substantially higher construction facilities for comfort cooling and thus have higher energy usage. The proportion of such areas in various buildings can range from 10% to 40% of the total volume [36] (pp. 815–816). The energy consumption may be comparable with the energy use in all other occupied building areas. Three times the electricity per unit area is used in these transitional spaces relative to the buildings' interior [37] (p. 633). So, the concept of comfort in a transitional space is very different from that of a well-regulated, air-conditioned workplace. Although the occupant's interaction with the thermal environment may be temporary, it may be proposed that broader-than-normal variations in thermal conditions should be studied. It may also be derived from current comfort standards that occupants can accommodate a broader spectrum of comfort in such circumstances [36] (p. 816).

Additionally, to define a mediated space, we should first determine the difference between outdoor and indoor spaces. Nicol and others define a number of factors that separate outdoor spaces from inside spaces [38] (pp. 1–2):

- Less human regulation than indoors: This would make the room less of a property of the person or community, and, consequently, their rights to change it are reduced.
- Varying climatic conditions: The seasonal weather conditions are higher outside than in, although this may be due to clothing or wind speed.
- Diversity of space: This is a function that will help people stay relaxed, offering relief from the sun and wind.
- Variety of use: For certain people, open areas are their workplace; for some, they are connected to recreation; and for others, they are a way of going from one indoor space to another. The multiple criteria and preferences will plan out the ideal environment differently.
- Wider comfort tolerance: The amount of discomfort currently calculated in outdoor environments is much lower than that projected for indoor spaces. This difference indicates that people could be more accepting of environments outdoors.
- Previous experience of space: While most people indoors have a clear sense of what to expect thermally, prior experience of the environment is less helpful in an outdoor space because the weather is continually changing, and that experience might be unusual [38] (pp. 1–2).

In Figure 5, four categories of criteria are introduced to understand and analyse mediated spaces in case studies discussed in the next section. These mediated space design criteria include community interactions, physical education activities, natural exposure, and outdoor/indoor informal learning. The hypothesis in this section is that, by using these criteria in mediated space design, well-being outcomes will be improved for children. These categories are the benchmarks that answer the following question: how can spaces mediate interaction between natural spaces and children's physical activity?

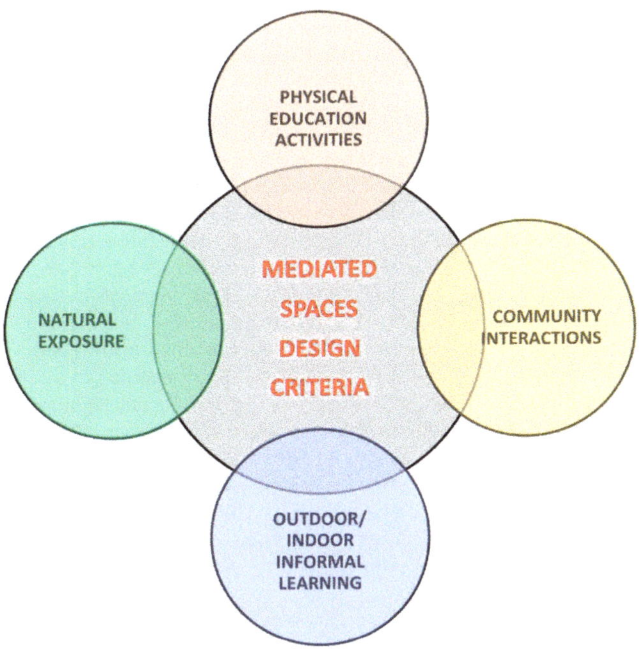

Figure 5. Categories for analysing and determining mediated spaces in case studies. Source: author.

5. Analysis of Mediated Spaces in Schools

In this section, six different case studies are evaluated concerning the identification of mediated spaces introduced in Section 4. The main criteria for identifying mediated spaces include community interactions, physical education activities, natural exposure, and outdoor/indoor informal learning. The reason behind choosing the following case studies is that each case study has a specific pedagogical model and related architectural concepts to exemplify the mediated space definition in this research. These case studies have also been chosen as they are well known for their best practices in learning environments in terms of using architecture to achieve pedagogical aspects by leading architects in the field of designing learning environments, such as Neutra and Hertzberger. Although this article focuses on vertical schools, there is another significant rationale for selecting a horizontal school case for this section, in addition to pedagogical reasons. As mentioned earlier, the absence of an established "rule book" for constructing vertical school projects leads to relying on horizontal practices and analysing the advantages and disadvantages of these designs [13]. The intention of analysing these case studies is to find mediated spaces in successful horizontal schools in various climates in order to adapt them in vertical schools in hot–humid climates using biophilic strategy considerations.

These case studies are as follows:

- Antonio Sant' Elia Kindergarten—Como, Italy—Giuseppe Terragni: 1937;
- Emerson Junior High School—Los Angeles, USA—Richard Neutra: 1938;
- UCLA Lab School—Los Angeles, USA—Neutra and Alexander: 1959;
- Montessori Primary School—Delft, Netherlands—Herman Hertzberger: 1960;
- Boarding School—Morella, Spain—Carme Pinós and Enric Miralles: 1994;
- Hellerup School—Copenhagen, Denmark—Arkitema: 2002.

Below, a summary of the research method is first presented. Then, some indicative floor plans of these case studies, along with an analysis of the interconnectivity of spaces, are included to assist in making connections between the types of spaces identified as mediated spaces. Following that, a brief summary is presented regarding what spaces could be considered as mediated spaces and what architectural elements have been used in mediated spaces for the above case studies according to the space syntax analysis.

Space syntax analysis is used to evaluate these case studies further. In this section, a summary of the theory and purpose of space syntax methodology and its components is first set out. Then, the selected approach is utilised for case study analysis.

The space syntax analysis developed by Hillier and his colleagues in 1976 [39] (p. 248) has grown into a worldwide community of researchers, practitioners, and consultants exploring a diverse array of phenomena. Its applications span beyond architecture and urban planning and include interdisciplinary contributions to fields like archaeology, history, sociology, management, neuroscience, and biology, among others [39] (pp. 248–249). In terms of architectural and urban design aspects, it has emerged as the most influential technique for studying street networks in relation to pedestrian movement. This method focuses on the configuration of spaces in buildings and towns, emphasising the arrangement and assembly of urban elements to create interconnected networks of connectivity [40] (p. 509). The central assumption of the space syntax theory is that space gains significance through the interconnection of its various components, forming a complex spatial system that people inhabit and navigate. By viewing space as a network of interconnected parts, such as rooms and corridors, the approach facilitates the measurement of a room or a corridor's importance in the entire system, like a building. The theory predicts usage frequency based on spatial centrality; highly integrated spaces see more activity, while more isolated spaces tend to be quieter [39] (p. 249).

Space syntax analysis is not a singular method but rather a cluster of approaches to analysing spatial relations—mainly gamma, isovist, and axial analyses (Figure 6). The three approaches to space syntax vary in both methodology and scale of application. Gamma analysis is predominantly confined to architectural scales, isovist analysis encompasses urban design scales as well, and axial analysis is applicable across all scales [40] (pp. 509, 512).

(a) (b) (c)

Figure 6. Different approaches in space syntax: (**a**) gamma; (**b**) isovist; (c) axial analysis. Source: [40] (p. 510).

"Gamma analysis" predominantly entails creating a diagrammatic map that illustrates the relationships between spatial segments within building interiors. In this approach, rooms and spaces are conceptualised as cells, with the building entry serving as a foundational point. In the context of gamma analysis, there is often a pursuit to recognise the morphogenic patterns or genotypes inherent in spatial structures that contribute to the creation of space in typical building types. The analysis of the depth and shallowness of spatial structures can provide insights into how buildings facilitate relationships between those within and outside the space [40] (pp. 508–510).

"Isovist mapping" is a technique used in space syntax to analyse the horizontal visual field from a specific point in space, representing the area visible without obstruction from walls or buildings. Essentially, it explores the visual experience of a given location and determines the scope of the visual field influenced by architectural and urban design elements. This includes considerations such as the impact of vehicles, trees, street signage, and furniture at short distances, as well as factors like topography, light, and air quality at longer distances. In practical terms, isovist mapping unveils locations with the most expansive visual fields, shedding light on how architecture influences privacy and shapes the perception of a visitor's gaze. The resulting isovist map quantifies the extent to which the broader spatial structure can be observed from a singular vantage point. It is important to note that while isovist extends through transparent boundaries, the technique does not measure physical access, as these visual fields cannot be physically traversed [40] (p. 511).

"Axial analysis" is a topological method that prioritises interconnectivity over distance considerations. This approach builds upon the concept of lines of sight by determining the minimum number of straight axes required to navigate through the network of public spaces. Each axis is treated as a singular element in the spatial network, irrespective of its length, and is assigned a single integration value. The axes with the highest integration are those that necessitate the fewest axial turns to connect with all other axes in the network [40] (p. 511).

According to what has been mentioned, as gamma analysis is largely limited to architectural scales and could represent the level of connectivity with all other cells within that building, this research focuses on gamma analysis. In this research, this method is used to evaluate the interconnectivity of spaces to see which spaces could be considered mediated spaces in the building. In addition, by constructing a convex map using space syntax, this research conducts a spatial configuration for two nearby classrooms and their adjacent public/outdoor spaces on one floor of each school.

As shown in the second column of Table 3, the "convex map" column, the relationship between two classrooms and their adjacent public/outdoor spaces from each case study is considered for the convex map analysis. Classrooms are identified with red colour, and the remaining areas are coloured dark blue. In Hellerup School, informal learning is taking place everywhere, including the central stairs, as the community/social interaction space connects all levels vertically with several sitting and nook areas surrounding it. It is interesting to note that in Hellerup School, with its open-plan typology, each indoor space can be regarded as a mediated space. This plan typology not only creates an informal learning space for children but also provides constant physical activities in the central Hellerup stairs, social interaction in the stair tread, and visual interaction in the void. The only category that is not considered in this school is the direct interaction of indoor spaces with nature and outdoors, which is completely located outside of the building and only used for formal breaks.

As can be seen from the third column of Table 3, the space syntax column, analysis often starts with each classroom point, which serves as the foundational point for spatial mapping. From this point, the connections to other spaces are traced, which are shown by red nodes. In addition, the number of circuits at each node demonstrates which spaces are likely to see higher levels of activity and which are more isolated. Highly integrated spaces are expected to be busier, while isolated spaces are quieter. The aim is to identify patterns that reveal how spaces are configured to facilitate movement, interaction, and accessibility. This is crucial for identifying mediated spaces that enhance learning and well-being.

For example, the gamma analysis map for Hellerup School reveals a highly integrated atrium that connects directly to most individual learning spaces and communal areas, indicating it is a central hub of activity. In Emerson Junior High School, the analysis shows a series of classrooms leading to multiple outdoor courtyards, suggesting easy access to natural spaces, which are crucial for student well-being. The map for UCLA Lab School highlights a deep spatial structure with multiple layers of spaces from the main entrance to classrooms, indicating various communal spaces but potential challenges in wayfinding and accessibility to nature. Boarding School highlights the connectivity of indoor spaces but notes the lack of direct access to outdoor spaces. Montessori Primary School demonstrates direct access to outdoor spaces from classrooms. In Antonio Sant'Elia Kindergarten, classroom access to indoor and outdoor spaces indicates integration and interaction potential.

Connecting classrooms to an outdoor space can be seen in all traditional classroom typologies, including Montessori Primary School, Emerson Junior High School, UCLA Lab School, and Antonio Sant' Elia Kindergarten. This outdoor learning space is sometimes exclusively used by one classroom but sometimes a common space between at least two classrooms, as can be seen at UCLA Lab School or Emerson Junior High School. Although two classes are connected directly to create a more interactive space for a larger group of students in Boarding School, there is no direct access to outdoor spaces in this case study. The classrooms of Montessori Primary School, Emerson Junior High School, UCLA Lab School, and Antonio Sant' Elia Kindergarten provide a direct connection to nature and outdoor space.

In most case studies, a hierarchy from indoor to outdoor spaces can be seen. Indoor spaces are mostly integrated into outdoor spaces through sheltered spaces like a terrace, porch, or staircase. The outdoor mediated spaces in this case study include sheltered porches and terraces, stairs, and internal courtyards, which are commonly used as informal learning spaces and for community interactions while providing natural exposure and scope for physical activity for the users. On the other hand, the rest of the mediated spaces include central staircases, students' communal kitchens, halls, and corridors. The shape of the corridors in most case studies is linear, while in Hellerup School, Boarding School, and Montessori Primary School, corridors are merged into a central hall space and create a rectangular shape.

Table 3. Spatial configuration analysis of mediated space design for each case study. Source: author.

School Name	Convex Map [1,2]	Graph of Space Syntax [3,4,5,6,7]	Mediated Space Relationship Diagram [8,9,10,11,12]
Antonio Sant'Elia Kindergarten Italy—Giuseppe Terragni: 1937	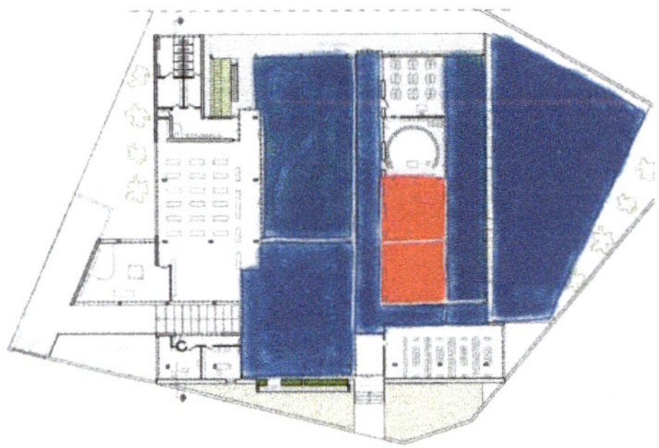	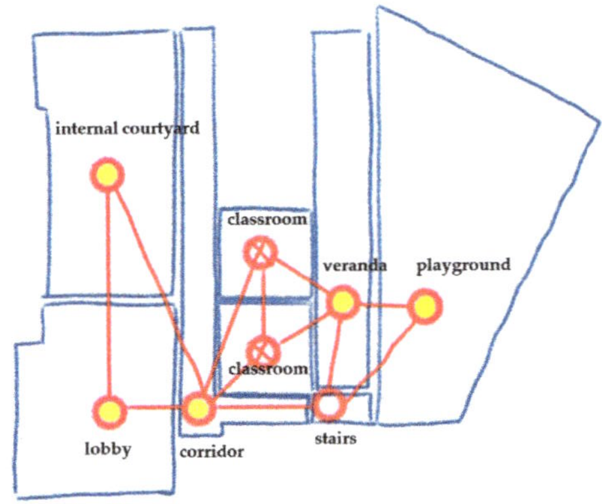	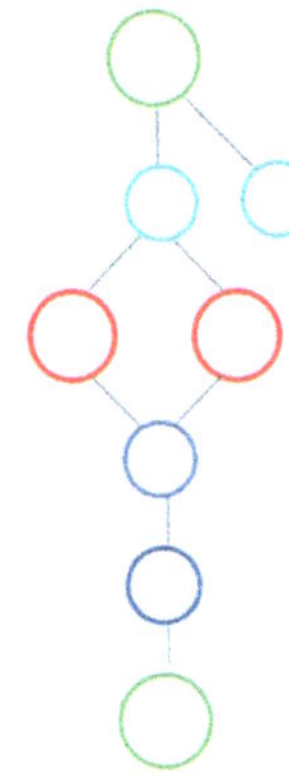
Emerson Junior High School USA—Richard Neutra: 1938	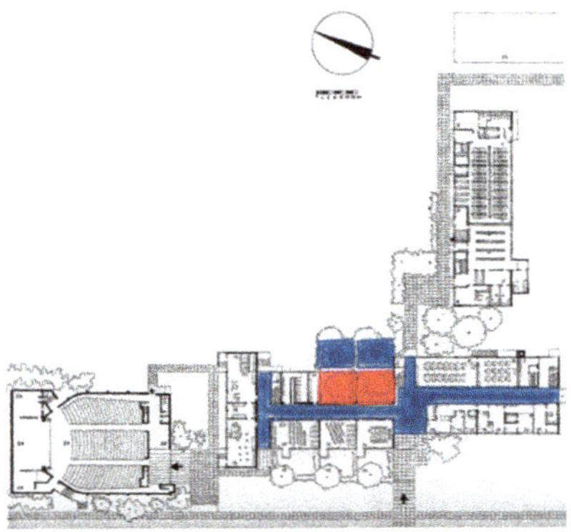	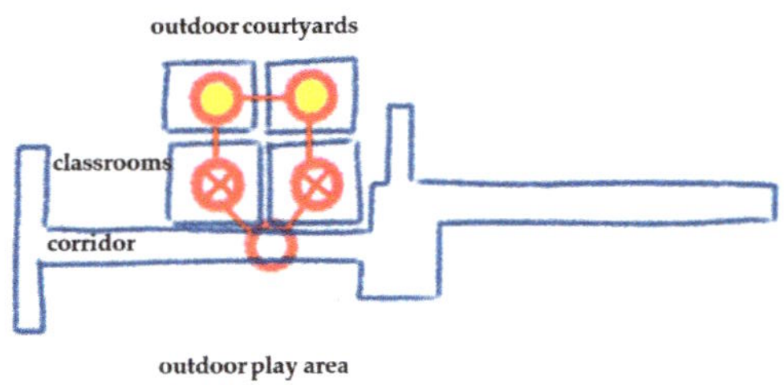	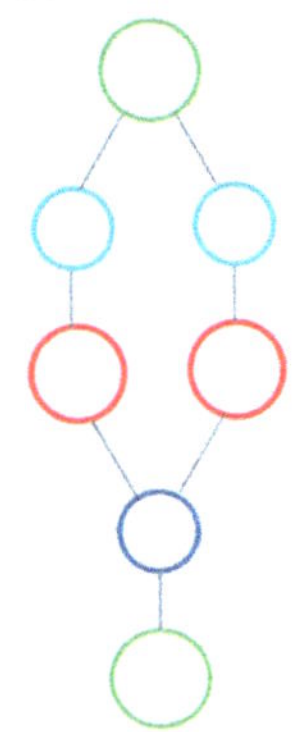

Table 3. *Cont.*

School Name	Convex Map [1,2]	Graph of Space Syntax [3,4,5,6,7]	Mediated Space Relationship Diagram [8,9,10,11,12]
UCLA Lab School USA—Neutra and Alexander: 1959	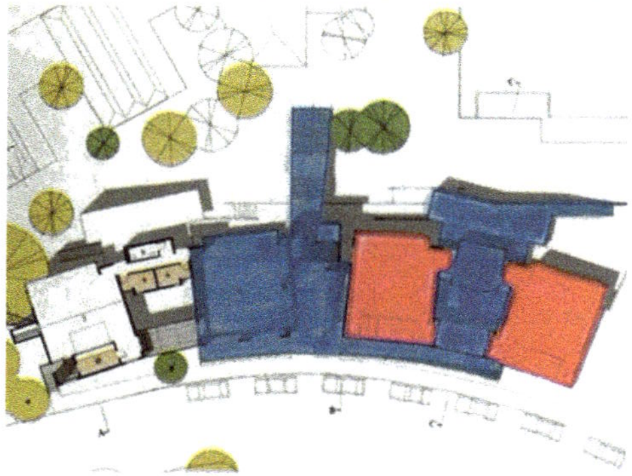	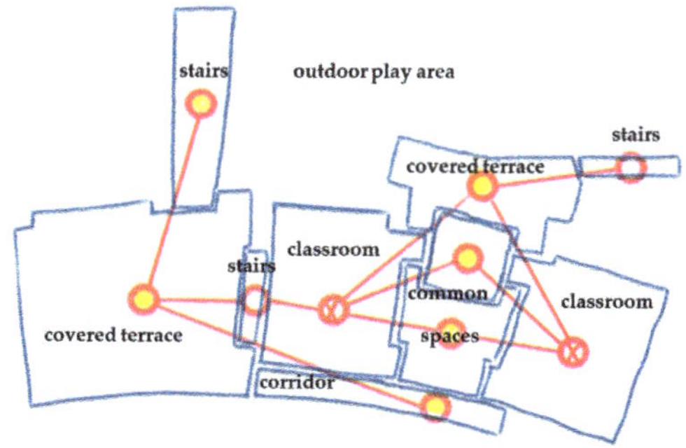	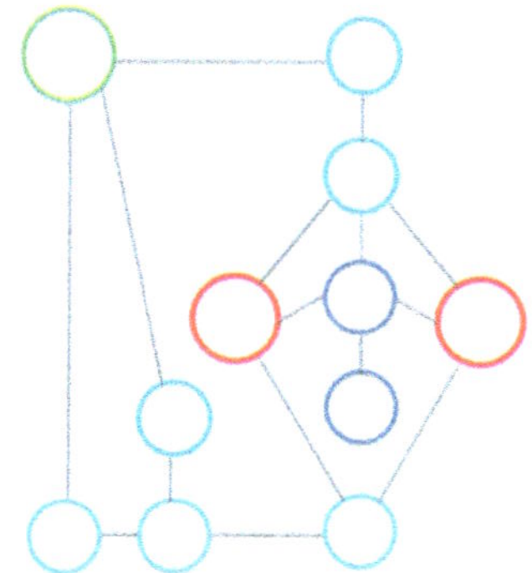
Montessori Primary School Netherlands—Herman Hertzberger: 1960	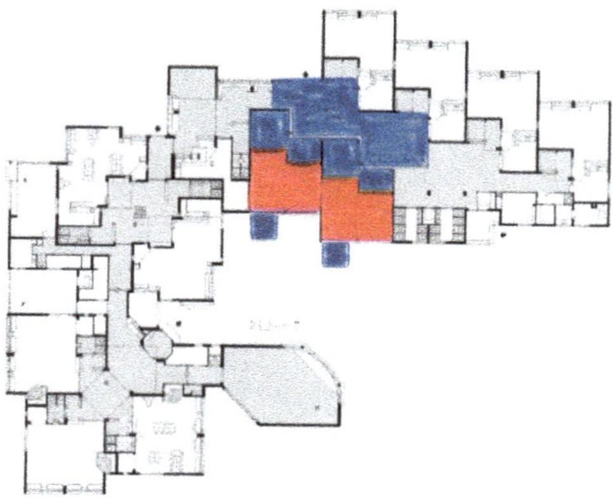	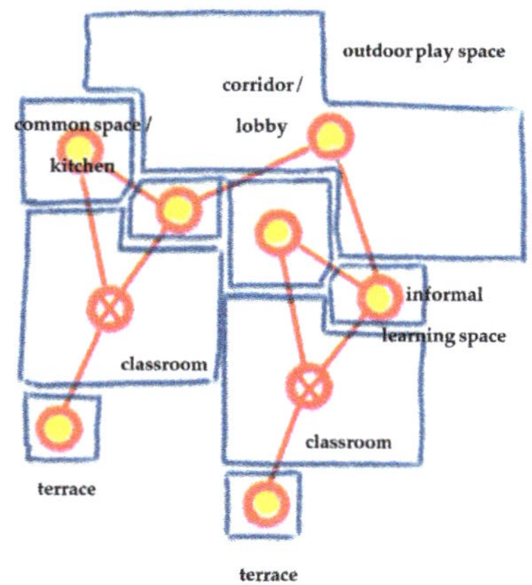	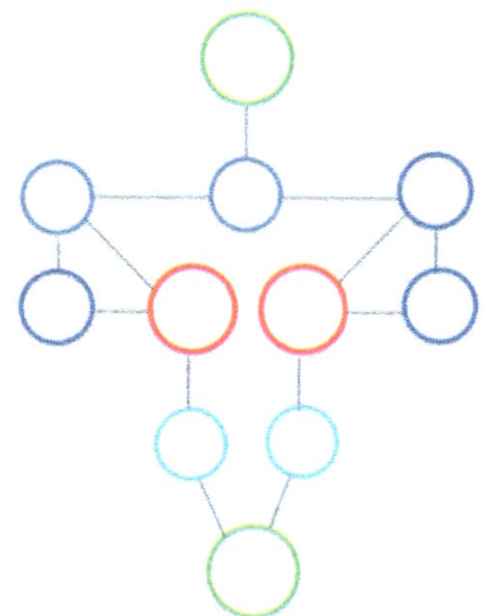

Table 3. *Cont.*

School Name	Convex Map [1,2]	Graph of Space Syntax [3,4,5,6,7]	Mediated Space Relationship Diagram [8,9,10,11,12]
Boarding School Spain—Carme Pinós and Enric Miralles: 1994	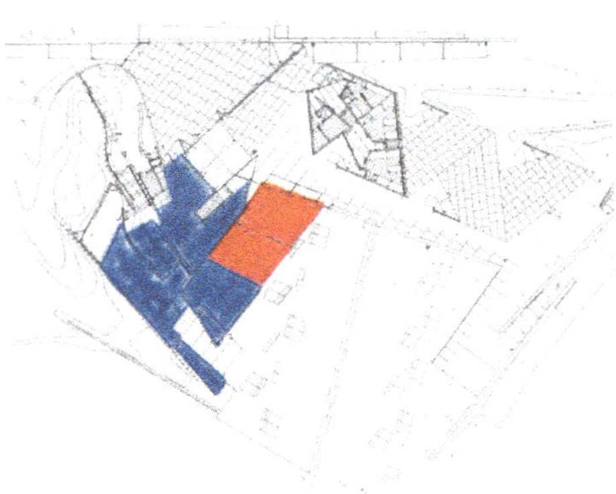	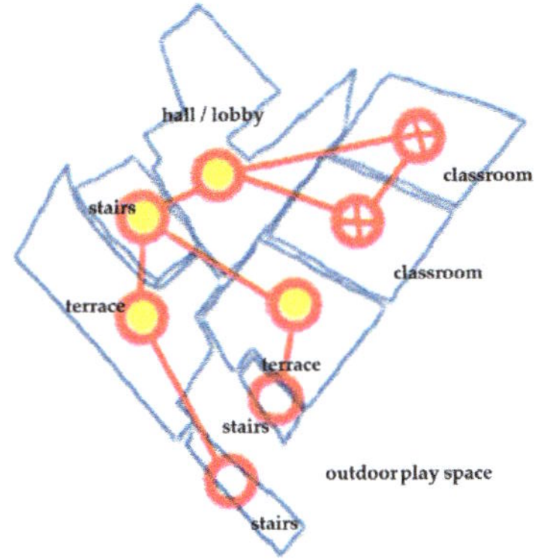	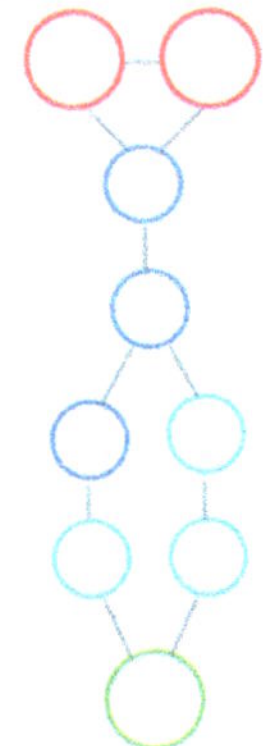
Hellerup School Denmark—Arkitema: 2002	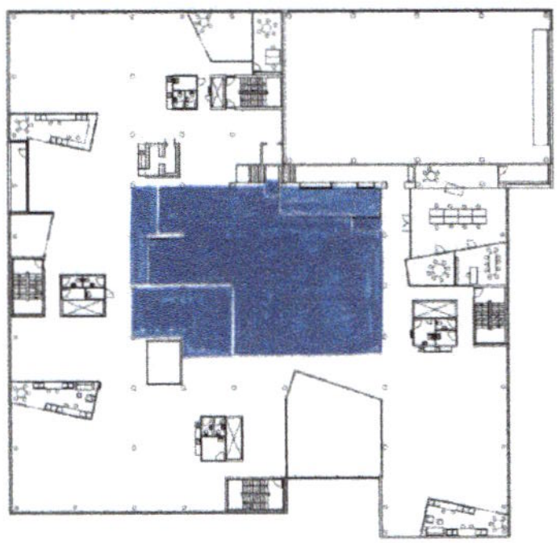	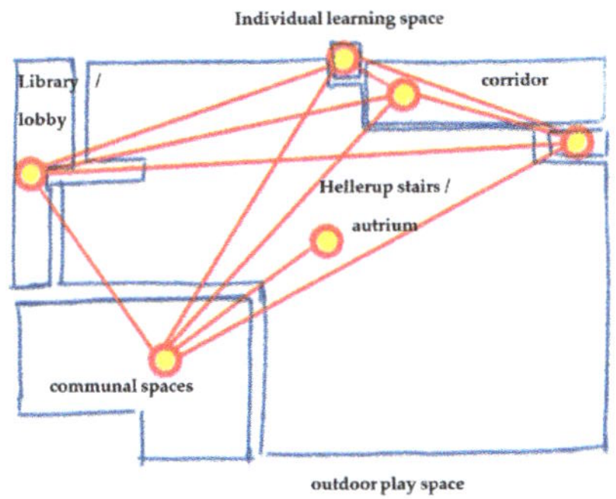	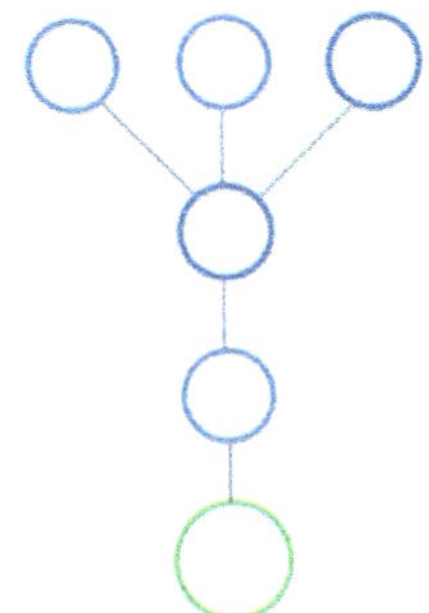

[1] Two nearby classrooms. [2] Adjacent public/outdoor spaces. [3] Most used adjacent spaces. [4] Less used adjacent spaces. [5] Classroom. [6] Circuits. [7] Occupied spaces. [8] Open space. [9] Classroom. [10] Mediated space (indoor). [11] Mediated space (outdoor). [12] Direct access.

The fourth column of Table 3, the mediated space relationship diagram column, is dedicated to the bubble diagram relationships between mediated spaces (indoor or outdoor), two adjacent classrooms, and open spaces of the six case studies. In each diagram, the hierarchy of indoor mediated spaces, outdoor mediated spaces, and open spaces according to two adjacent classrooms is illustrated. Except for Hellerup School, which is entirely an open-plan school without standard traditional classrooms, all other schools have typical classrooms. Only Boarding School classrooms have no direct access to outdoor mediated spaces, while in other case studies, classrooms have direct access to both indoor and outdoor mediated spaces.

On the other hand, in all case studies, open space access is located at the far end of the diagram compared to the position of classrooms. In Hellerup School, indoor mediated spaces are connected to the open space without the existence of outdoor mediated spaces. In Boarding School and UCLA Lab School, classroom access to the open spaces is through outdoor mediated spaces. In contrast, in Montessori Primary School, Emerson Junior High School, and Antonio Sant' Elia Kindergarten, classrooms have access to open spaces through both outdoor and indoor mediated spaces.

In all the school case studies, except Hellerup School, which is an open-plan school and does not include classrooms at all, and Boarding School, classrooms are located at the centre of the hierarchy diagrams, and open spaces are at the end of each hierarchy diagram. Access to the open spaces in Montessori Primary School, Emerson Junior High School, and Antonio Sant' Elia Kindergarten is from two directions, which promotes natural and outdoor interaction in schools.

The following parameters are the authors' attempt to construct an analytical method and system of criteria for identifying and designing mediated spaces in vertical schools. As can be seen from Table 4, the case studies cover nearly all required criteria for designing mediated spaces (three to four categories) with various architectural elements and spaces. Among the architectural elements of mediated spaces are the areas of stairs, nook areas, common learning spaces, internal courtyards/gardens, lobbies, communal kitchens, terraces, and corridors. It is interesting to note that Hellerup School, the case study with open-plan typology, does not have a distinct border between formal and informal learning spaces, and the majority of its spaces can be considered mediated spaces. Although community interaction, physical activity, and informal learning spaces are considered in this case study, a lack of direct interaction between learning spaces and the natural environment can be seen. However, Antonio Sant' Elia Kindergarten provides both internal and external courtyard spaces that connect indoor spaces to the surrounding natural environments. As can be seen from the table, except for the open-plan Hellerup School, other case studies include traditional classrooms. However, traditional school plan typologies can have direct natural exposure within the school building.

Table 4. Spaces and architectural elements demonstrating mediated space design criteria. Source: author.

Case Studies	Criteria	Elements and Spaces
Antonio Sant' Elia Kindergarten Como, Italy—1937 Giuseppe Terragni	Community/social interaction Natural exposure Outdoor/indoor informal learning	Lobby Corridor Internal courtyard Outdoor courtyard Garden Veranda with canopy Transparent façade

Table 4. *Cont.*

Case Studies	Criteria	Elements and Spaces
Emerson Junior High School Los Angeles, USA—1938 Richard Neutra	Community/social interaction Physical education activity Natural exposure Outdoor/indoor informal learning	Garden Courtyard Corridor Lobby Auditorium Sliding glass doors/walls Transparent façade Trees Outdoor classroom
UCLA Lab School Los Angeles, USA—1959 Richard Neutra and Robert Alexander	Community/social interaction Physical education activity Natural exposure Outdoor/indoor informal learning	Children's garden Common learning space Terrace Porch Stairs/bridge Corridor Outdoor courtyard Trees Lawn Transparent façade Sliding window walls In-between spaces like porch and patio
Montessori Primary School Delft, Netherlands—1960 Herman Hertzberger	Community/social interaction Physical education activity Natural exposure Outdoor/indoor informal learning	Terrace Corridor library with chimney Communal kitchen Non-enclosed outdoor courtyard/street space Children's garden Direct access to outside from the classroom terrace Transparent large façade Window seating area
Boarding School Morella, Spain—1994 Enric Miralles and Carme Pinós	Community/social interaction Physical education activity Outdoor/indoor informal learning	Stairs Hall/lobby Terrace Corridor Patio Openable apertures/pivot doors Outdoor courtyard Game court Transparent façade/window walls
Hellerup School Copenhagen, Denmark—2002 Arkitema Architects	Community/social interaction Physical education activity Outdoor/indoor informal learning	Atrium Hellerup stairs Outdoor courtyard Gymnasium Nook spaces like balconies and bridges

It is argued that these case studies have been chosen because they demonstrate pedagogical and socialising design principles as well as mediated relationships with the natural world. However, for these principles to be applied in a hot–humid climate and high-rise context, considering biophilic principles like the use of natural elements of green features in these spaces, they would need to be adapted in the following ways:

- Vertical gardens and green walls: Horizontal pedestrian-orientated gardens can be adapted into vertical gardens and green walls with low-maintenance plants, creating opportunities for natural exposure and informal learning in vertical schools.
- Sky terraces and rooftops: Outdoor courtyards and play areas can be transformed into sky terraces and rooftops, providing space for outdoor activities and sports.
- Bridges and ramps: Traditional courtyards can become habitable link bridges and ramps, maintaining connectivity and providing additional space for movement and social interaction.
- Climbing spaces: Adventure spaces in horizontal schools can be converted into climbing spaces in vertical schools, promoting physical activity.
- Internally enclosed courtyards: Non-enclosed outdoor courtyards or street spaces in horizontal schools can be adapted into internally enclosed courtyards connecting multiple levels, utilising staircases or slides for movement.

These adaptations ensure that vertical schools can maintain or even enhance the functionality and benefits of spaces traditionally found in horizontal schools.

6. Results: Well-Being Benchmarks/Criteria of Mediated Space Design

In this section, the elements required for designing mediated spaces that result in maintaining well-being are presented. These criteria are extracted from the definition of mediated spaces in Section 4, Figure 5. Then, the corresponding elements and spaces needed to achieve such criteria in design are presented in the design guidelines columns. These are extracted from the analysis of the case studies of successful learning environments with mediated space design in Section 5 and the biophilic design principles in Section 3.4.

As presented in Table 5, the following section explains each criterion for the design guidelines:

- Criterion 1—Flexible Seating: Provide flexible seating arrangements for students' social interaction, including studying together, eating, and mingling with other students from higher levels.
- Criterion 2—Vibrant Space: Provide gallery spaces and wall hangings to create a more vibrant space for socialising.
- Criterion 3—Visual Interaction: Provide transparency, visual, and natural interaction by creating views over voids, skylight/roof windows, and glass facades/walls/doors into the semi-outdoor or outdoor spaces.
- Criterion 4—Greening Strategies: Use of indoor plants, small gardening boxes, grass in semi-outdoor spaces and trees, vegetable gardens, and green walls in outdoor mediated or semi-outdoor conservatory spaces.
- Criterion 5—Extendable Spaces: Extend spaces outside by incorporating pivoting doors/windows and sliding glass walls/windows, i.e., provide capacity for spaces to adapt to variations in climatic conditions.
- Criterion 6—Personal Spaces: Create personal spaces for students while still remaining under the supervision of school staff.
- Criterion 7—Shading Solutions: Increase the extent of shading using cool surface materials, planting trees with dense canopies, turf wicking, and establishing green roofs and walls.
- Criterion 8—Natural Interaction: Provide interaction with nature both visually and by using natural materials like wood and elements like rocks, sand, and fire inside and outside the building.

Table 5. General design guidelines for mediated spaces. Source: author.

Mediated Spaces \ Design Guidelines/Criteria	1 Flexible Seating	2 Vibrant Space	3 Visual Interaction	4 Greening Strategies	5 Extendable Spaces	6 Personal Spaces	7 Shading Solutions	8 Natural Interaction
Lobby	✓	✓	✓	✓	✓	✓	NA [1]	✓
Atrium	✓	✓	✓	✓	✓	NA	✓	✓
Corridor	✓	✓	✓	✓	✓	✓	NA	✓
Sky bridges	✓	✓	✓	✓	NA	NA	NA	✓
Library	✓	NA	✓	✓	✓	✓	NA	✓
Plaza	✓	✓	✓	NA	✓	NA	NA	✓
Multi-purpose hall	NA	✓	✓	NA	✓	NA	NA	✓
Auditorium	NA	NA	✓	NA	✓	NA	NA	✓
Communal kitchen	✓	NA	✓	NA	NA	NA	NA	✓
Canteen	✓	NA	✓	✓	✓	✓	NA	✓
Lunchroom	✓	NA	✓	✓	✓	✓	NA	✓
Staircases with void	NA	✓	✓	✓	NA	NA	NA	✓
Hellerup stairs	NA	✓	✓	✓	✓	✓	NA	✓
Multi-purpose hall	NA	NA	✓	NA	NA	NA	NA	✓
Sheltered amphitheatre	NA	NA	✓	✓	✓	NA	✓	✓
Window seat/nook area	NA	NA	✓	✓	NA	✓	NA	✓
Wildflower garden	NA	NA	NA	✓	NA	NA	✓	✓
Edible/food garden	NA	NA	NA	✓	NA	NA	✓	✓
Science/pollinator garden	NA	NA	NA	✓	NA	NA	✓	✓
Roof/sky garden	✓	NA	✓	✓	NA	✓	✓	✓
Pond	NA	NA	NA	NA	NA	NA	✓	✓
Sandpit	NA	NA	NA	NA	NA	NA	✓	✓
Nature Trail	✓	NA	NA	✓	NA	NA	✓	✓
Walkway path	✓	NA	NA	✓	NA	NA	✓	✓

Table 5. *Cont.*

Design Guidelines/ Criteria / Mediated Spaces	1 Flexible Seating	2 Vibrant Space	3 Visual Interaction	4 Greening Strategies	5 Extendable Spaces	6 Personal Spaces	7 Shading Solutions	8 Natural Interaction
Natural terrain	✓	NA	NA	✓	NA	NA	✓	✓
Grassy berm	✓	NA	NA	✓	NA	NA	✓	✓
Lawn	✓	NA	NA	✓	NA	NA	✓	✓
Outdoor classroom	✓	✓	✓	✓	✓	✓	✓	✓
Common learning space	✓	✓	✓	✓	✓	✓	✓	✓
Internal courtyard	✓	NA	✓	✓	✓	✓	✓	✓
Internal play area	✓	NA	✓	✓	✓	✓	✓	✓
Terrace	✓	NA	✓	✓	✓	✓	✓	✓
Balcony	✓	NA	✓	✓	✓	✓	✓	✓
Veranda with canopy	✓	NA	✓	✓	✓	✓	✓	✓
Pool	✓	NA	NA	NA	NA	NA	NA	✓
Gym	✓	NA	✓	NA	NA	NA	✓	✓
Climbing wall	NA	NA	✓	NA	NA	NA	✓	✓
Incline mound space	NA	NA	✓	NA	NA	NA	✓	✓
Outdoor courtyard	✓	NA	✓	✓	✓	✓	✓	✓
External play space	✓	NA	✓	✓	✓	✓	✓	✓
Rooftop play ground	✓	NA	✓	✓	NA	✓	✓	✓
Grass sport pitches	✓	NA	NA	✓	NA	NA	✓	✓
Hard game court	✓	NA	NA	✓	NA	NA	✓	✓
Running tracks	NA	NA	NA	✓	NA	NA	✓	✓

Community/social interactions . Natural exposure . Outdoor/indoor informal learning . Physical education activities . [1] NA: not applicable.

7. Conclusions

This article has explored the dynamic relationship between educational theory, environmental design, and the integration of nature in educational spaces, focusing on the concept of mediated spaces. It has discussed the transformative nature of learning and emphasised the importance of meeting students' basic needs. Theoretical perspectives from educationalists such as Dewey, Montessori, Piaget, Malaguzzi, and Vygotsky offer insights into various dimensions of children's development, emphasising the role of experience, movement, social interaction, and the environment in cognitive and social growth. The analysis covered various educational theories and pedagogical approaches, contrasting traditional methods with student-centred learning. In addition, by incorporating natural elements, daylight, and vegetation into building design, biophilic approaches can foster connections with nature, improve air quality, and support cognitive and emotional well-being.

The emergence of vertical schools presents both challenges and opportunities, particularly regarding access to nature and outdoor spaces. Therefore, mediated spaces were proposed as an alternative solution to enhance well-being and resilience in vertical school environments. Mediated spaces serve as critical interfaces between indoor and outdoor environments, facilitating informal learning, physical activity, and social interaction.

The study has evaluated six case studies, each representing a distinct pedagogical model and architectural approach, serving as exemplars of successful learning environments. Space syntax analysis was employed to evaluate the interconnectivity of spaces within school buildings, emphasising the importance of spatial configuration in promoting engagement and movement. Common themes across the case studies included the integration of classrooms with outdoor spaces, central staircases as hubs for social interaction, and the incorporation of natural elements.

The findings have led to the proposal of design guidelines aimed at creating holistic, adaptive educational environments that support diverse learning needs and promote well-being. Overall, the synthesis of learning theory and environmental design underscores the significance of architectural design in shaping educational experiences and fostering student development.

Author Contributions: Conceptualisation, E.E.S.; methodology, E.E.S., N.W.; validation, E.E.S., N.W.; formal analysis, E.E.S.; investigation, E.E.S.; resources, E.E.S.; data curation, E.E.S., N.W.; writing—original draft preparation, E.E.S.; writing—review and editing, E.E.S., N.W.; visualisation, E.E.S.; supervision, N.W.; All authors have read and agreed to the published version of the manuscript.

Funding: This research received no external funding.

Institutional Review Board Statement: Not applicable.

Informed Consent Statement: Not applicable.

Data Availability Statement: No new data were created or analyzed in this study. Data sharing is not applicable to this article.

Conflicts of Interest: Authors declare no conflict of interest.

References

1. Ariès, P. *Centuries of Childhood: A Social History of Family Life*; Vintage Books/Random House: New York, NY, USA, 1962.
2. Landvoigt, C. *John Dewey's Concept of the Child*; ProQuest Dissertations Publishing: Ann Arbor, MI, USA, 1986. Available online: https://www.proquest.com/dissertations-theses/john-deweys-concept-child/docview/303444562/se-2?accountid=14681 (accessed on 13 July 2024).
3. Marín Murcia, J.P.; Martínez Ruiz-Funes, M.J. Froebel and the teaching of botany: The garden in the Kindergarten Model School of Madrid. *Paedagogica Historica.* **2020**, *56*, 200–216. [CrossRef]
4. Schutte, A.R.; Torquati, J.C.; Beattie, H.L. *Impact of Urban Nature on Executive Functioning in Early and Middle Childhood. Environment and Behavior*; SAGE Publications: Los Angeles, CA, USA, 2017; Volume 49, pp. 3–30. [CrossRef]
5. Duffy, A. The 'NATURE' of New School Design—An Evolving Concept; Australia. 2019. Available online: https://architectus.com.au/insight/the-nature-of-new-school-design-an-evolving-concept/ (accessed on 11 January 2021).
6. Kim, D. *Vertical School: A New Typology for the City*; ProQuest Dissertations Publishing: Ann Arbor, MI, USA, 2017. Available online: https://www.proquest.com/dissertations-theses/vertical-school-new-typology-city/docview/2015118063/se-2?accountid=14681 (accessed on 13 July 2024).
7. OWP/P Architects Inc. *VS Furniture, and Bruce Mau Design. The Third Teacher: 79 Ways You Can Use Design to Transform Teaching & Learning*; Abrams: New York, NY, USA, 2010.
8. *Connecting with Others*; Department of Health and Aged Care, Government of Australia: Canberra, Australia, 2019. Available online: https://www.headtohealth.gov.au/living-well/connecting-with-others (accessed on 27 December 2022).

9. Martin, T. *How Urban Planning Can Help Us Cope with Climate Change*; ArchDaily: Zurich, Switzerland, 2019. Available online: https://www.archdaily.com/909476/how-urban-planning-can-help-us-cope-with-climate-change (accessed on 18 January 2019).
10. Newton, C. Vertical schools on the rise in Australian cities. *Archit. Aust.* **2019**, *108*, 86–91. Available online: https://architectureau.com/articles/vertical-schools-on-the-rise/ (accessed on 27 December 2022).
11. Matthews, T.; Newton, C.; Guaralda, M.; Mayere, S. Vertical Shool as Community Hubs. In *Building Connections for Community Benefit*; Cleveland, B., Chandler, P., Backhouse, S., Clinton, J., McShane, I., Newton, C., Eds.; Proceedings of Schools as Community Hubs International: Melbourne, Australia, 2020. [CrossRef]
12. Swinburn, A. Vertical School Design: Strategising the spatial configuration of a multi-storey typology to facilitate education in dense city environments. In *Byera Hadley Travelling Scholarships Journal Series*; NSW Architects Registration Board: Sydney, Australia, 2017. Available online: https://www.architects.nsw.gov.au/resource-library/research-reports/report/73 (accessed on 31 May 2019).
13. *The Early Learning Village*; Bogle Architects: London, UK, 2017. Available online: https://www.boglearchitects.com/early-learning-village-singapore-1 (accessed on 29 March 2021).
14. Ebrahimi Salari, E.; Tenorio, R.; Westbrook, N. Vertical Schools and Mediated Spaces: The Necessity of Interaction with Natural Environment. In *Teaching-Learning-Research: Design and Environments*; Laura Sanderson, S.S., Ed.; Routledge: Manchester, UK, 2020; Volume 22, pp. 145–156.
15. Leatherbarrow, D. *Uncommon Ground: Architecture, Technology, and Topography*; MIT Press: Cambridge, MA, USA, 2000. [CrossRef]
16. Ambrose, S.A.; Bridges, M.W.; DiPietro, M. *How Learning Works Seven Research-Based Principles for Smart Teaching*; Wiley: Hoboken, NJ, USA, 2010.
17. Tetsutani, S. *Design Elements for a Vertical School: Case Study: Kaka'ako*; ProQuest Dissertations Publishing: Ann Arbor, MI, USA, 2018. Available online: https://www.proquest.com/dissertations-theses/design-elements-vertical-school-case-study/docview/2113513282/se-2?accountid=14681 (accessed on 13 July 2024).
18. Wardle, F. *Approaches to Early Childhood and Elementary Education*; Nova Science Publishers, Inc.: Hauppauge, NY, USA, 2009.
19. Dewey, J. My Pedagogic Creed. *J. Educ.* **1926**, *104*, 542. [CrossRef]
20. Machemer, P.L.; Crawford, P. Student perceptions of active learning in a large cross-disciplinary classroom. *Act. Learn. High. Educ.* **2007**, *8*, 9–30. [CrossRef]
21. Dovey, K.; Fisher, K. Designing for adaptation: The school as socio-spatial assemblage. *J. Archit.* **2014**, *19*, 43–63. [CrossRef]
22. Imms, W.; Mahat, M.; Byers, T.; Murphy, D. *Type and Use of Innovative Learning Environments in Australasian Schools: ILETC Survey No. 1*; The University of Melbourne: Melbourne, Australia, 2017. Available online: http://www.iletc.com.au/publications/reports/ (accessed on 21 June 2021).
23. *Children's Time to Play: A Literature Review*; NCB: London, UK, 2009. Available online: https://www.playday.org.uk/campaigns-3/previous-campaigns/2009-make-time/2009-reseach/ (accessed on 18 July 2024).
24. Dowdell, K.; Gray, T.; Malone, K. Nature and its influence on children's outdoor play. *Aust. J. Outdoor Educ.* **2011**, *15*, 24–35. [CrossRef]
25. Chawla, L. Benefits of Nature Contact for Children. *J. Plan. Lit.* **2015**, *30*, 433–452. [CrossRef]
26. Armitage, M. The influence of school architecture and design on the outdoor play experience within the primary school. *Paedagog. Hist.* **2005**, *41*, 535–553. [CrossRef]
27. Sack, M. *Richard Neutra*; In Studiopaperback Series; Verlag für Architektur: Zürich, Switzerland, 1992.
28. Carroli, L.; Willis, J.; Franz, J.; Barker, S.; Bone, N.; Brown, A.; Chisholm, N.; Gibson, A.; Gillett-Swan, J.; Harth, S.; et al. *Conversations about Vertical Schools in Australia*; QUT: Brisbane, Australia, 2022. Available online: https://research.qut.edu.au/tvs/2023/04/18/broadening-the-conversation-tvs-launches-discussion-paper/ (accessed on 16 October 2023).
29. Moehring, J. *The 21st Century Learning Environment: Education in the Digital Era*; ProQuest Dissertations Publishing: Ann Arbor, MI, USA, 2012. Available online: https://www.proquest.com/dissertations-theses/21st-century-learning-environment-education/docview/1312738818/se-2?accountid=14681 (accessed on 13 July 2024).
30. Truong, S.; Singh, M.; Reid, C.; Gray, T.; Ward, K. Vertical schooling and learning transformations in curriculum research: Points and counterpoints in outdoor education and sustainability. *Curric. Perspect.* **2018**, *38*, 181–186. [CrossRef]
31. Beatley, T.; Newman, P. Biophilic Cities Are Sustainable, Resilient Cities. *Sustainability* **2013**, *5*, 3328–3345. [CrossRef]
32. Wilson, E.O. *Biophilia*; Harvard University Press: Cambridge, MA, USA, 1984. [CrossRef]
33. Newman, P. Biophilic urbanism: A case study on Singapore. *Aust. Plan.* **2014**, *51*, 47–65. [CrossRef]
34. *Better Placed Environmental Design in Schools: Making Schools Feel and Work Better*; Department of Education, Government Architect NSW: Sydney, Australia, 2018.
35. Can, I.; Heath, T. In-between spaces and social interaction: A morphological analysis of Izmir using space syntax. *J. Hous. Built Environ.* **2016**, *31*, 31–49. [CrossRef]
36. Pitts, A.; Saleh, J.B. Potential for energy saving in building transition spaces. *Energy Build.* **2007**, *39*, 815–822. [CrossRef]
37. Chun, C.; Tamura, A. Thermal comfort in urban transitional spaces. *Build. Environ.* **2005**, *40*, 633–639. [CrossRef]
38. Nicol, F.; Wilson, E.; Ueberjahn-Tritta, A.; Nanayakkara, L. *Comfort in outdoor spaces in Manchester and Lewes, UK*; Oxford Brookes University: Oxford, UK, 2006.

39. Kerstin Sailer, P.K. Space Syntax Theory: Understanding human movement, co-presence and encounters in relation to the spatial structure of workplaces. In *A Handbook of Theories on Designing Alignment between People and the Office Environment*; Routledge: Manchester, UK, 2021; pp. 248–260.
40. Pafka, E.; Dovey, K.; Aschwanden, G.D.P.A. Limits of space syntax for urban design: Axiality, scale and sinuosity. *Environ. Plan. B Urban Anal. City Sci.* **2020**, *47*, 508–522. [CrossRef]

MDPI AG
Grosspeteranlage 5
4052 Basel
Switzerland
Tel.: +41 61 683 77 34

Architecture Editorial Office
E-mail: architecture@mdpi.com
www.mdpi.com/journal/architecture

www.ingramcontent.com/pod-product-compliance
Lightning Source LLC
Chambersburg PA
CBHW061511210926
51850CB00008B/1877